DEATH AT BUFFALO CREEK

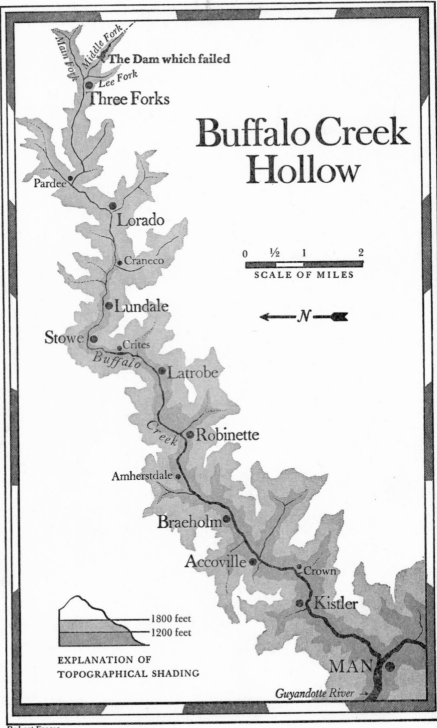

Main Fork

Middle Fork

Lee Fork

The Dam which failed

Three Forks

Buffalo Creek Hollow

Pardee

Lorado

Craneco

0 ½ 1 2

SCALE OF MILES

Lundale

←— N —←

Stowe

Crites

Buffalo

Latrobe

Creek

Robinette

Amherstdale

Braeholm

Accoville

Crown

Kistler

1800 feet

1200 feet

MAN

**EXPLANATION OF
TOPOGRAPHICAL SHADING**

Guyandotte River →

Robert Freese

DEATH
at
Buffalo Creek

The 1972
West Virginia Flood
Disaster

TOM NUGENT

W·W·NORTON & COMPANY·INC·
New York

COPYRIGHT © 1973 BY W. W. NORTON & COMPANY, INC.

First Edition

Library of Congress Cataloging in Publication Data

Nugent, Tom.
  Death at Buffalo Creek; the 1972 West Virginia flood
disaster.

  1. Buffalo Creek, W. Va.—Flood, 1972. 2. Dam
failures. I. Title.
TC557.W42B846 1973   975.4′44′04   73–3328
ISBN 0–393–05482–9

Some of the material in this book originally appeared,
in somewhat different form, in the *Detroit Free Press*
and is reprinted here with their permission.

PRINTED IN THE UNITED STATES OF AMERICA

1 2 3 4 5 6 7 8 9 0

T O

The People of Buffalo Creek

CONTENTS

PREFACE

THE VOICE rose and fell, a high-pitched, wailing sound, and the words went floating out to the deep green hills of southern West Virginia:

> *Listen today to a story I tell,*
> *In a sad and teardom way;*
> *Of a dreadful mudpile that the company turned*
> *   loose,*
> *And washed our people away.*

The man sat on a tilted, wood-slat porch in the middle of July, his old fingers busy on the neck of his guitar, his gaze traveling now and then to the summergreen mountains above him. The voice rose and fell: floated past the fat hound who lay deadsleep in the shadow of the weeping willow tree; past the grunting snuffle of the hog pen; past the rows of greenripe cabbage and corn, to dissolve at last in the droning silence of the creek bottom.

> *Our mothers so dear, and our fathers the same,*
> *Washed down that terrible stream;*
> *Searchin' and cryin', we found 'em all there,*
> *They died with poison and debris.*

The man's name was Hatfield, Wayne Brady Hatfield, he'd been a coal miner all his life, he'd lost everything in the world he loved, and now all he wanted was to see justice

9

done. "Wherever it takes for me to go, I'll go," he said. "Whatever has to be done, I'll do it. If it takes me the rest of my life, I'm gonna get justice done." Wayne Hatfield, mountaineer, coal miner, descendant of "Devil Anse" Hatfield and all his kin, descended from the furious, smoldering days of Hatfield-McCoy—now he sat on the wood-slat porch, massive and outraged, with no place to take his anger. He sang:

Now the company can kill whoever they please:
Away with the people—our insurance will pay!
Back home, back home, back down on Buffalo Creek . . .
Now it's so sad and alone.

Wayne Hatfield wrote his "Ballad of Buffalo Creek" in the summer of 1972—five months after a devastating flash flood which killed 125 people and destroyed or damaged more than 1,000 homes at a place called Buffalo Creek Hollow, a twisting, seventeen-mile-long creek bottom located fifty miles south of Charleston, West Virginia. The flood had taken Hatfield's wife of thirty-three years, Etta Pearl; it had taken his twenty-five-year-old daughter, Judith Annette, and his baby granddaughter, Connie Sue.

The five-bedroom, green-roofed home for which Hatfield had spent a lifetime working in the coal mines was gone, along with his automobile and his tractor. Disabled since 1959 with a crippling case of miners' black lung, Hatfield himself was slowly choking to death on the coal dust which he'd been breathing for thirty years. Justice was about all he had to live for now, and Hatfield trembled with a racing, barely controlled rage as he talked about the big coal company responsible for the flood. "They killed all them people," he said, "just the same as if they'd shot 'em. Anybody who would build a dam like that, and not take care of it right, and then let it break on

him without warning a soul . . . why, that man should
be tried and hung just like a criminal. It was murder, that's
what I say."

For Hatfield and his fellow coal miners—described by
sociologists as "the most systematically exploited and
expendable class of citizens in the nation"—the Buffalo
Creek disaster was only the latest in a seemingly endless
series of coal-related catastrophes. Most of them remembered November of 1968, for example, when seventy-eight
miners perished in a terrible underground explosion at
Consolidation's No. 9 mine near Farmington, West Virginia.
They remembered Christmas of 1951, when another 119
men were killed in an explosion at West Frankfort, Illinois.
And the old timers could go all the way back to the worst
disaster of them all—a 1907 explosion at Monongah, in the
northern part of the state, which killed 361.

The big disasters made headlines, but Hatfield and his
neighbors knew that most coal-mining deaths and injuries
came in individual accidents—when a miner was suddenly
buried under tons of falling "slate," or caught his hand in
a piece of heavy machinery. The evidence on safety could
not be disputed: with more than 100,000 deaths and 1 million injuries on the record books in its one hundred-year
history, coal mining was easily the most dangerous occupation in the United States.

Often, the American coal miner had been passive in the
face of these gruesome statistics—either unwilling or
unable to seriously challenge the big companies which
have traditionally controlled both the job market and the
politics in Appalachian states like West Virginia. After
Buffalo Creek, however, there was increasing evidence
that the mood had begun to change:

"The people of Logan County and West Virginia must
unite," said Don Bryant, chairman of the local Black Lung
Association (a group formed to lobby for better miner

disability benefits), at a hastily called citizens meeting two weeks after the flood. "Together, they must move step by step. They must decide that they aren't going to be dominated by the coal operators any more; that they aren't going to sit around waiting for the next disaster." There were rousing cheers for this, and shouts of "Nixon's a thief, too!" and "This wouldn't have happened if John L. Lewis was still running the UMW!" and "We're gonna throw Arch Moore [the Republican governor of West Virginia, often identified with coal interests] out of office!"

It had been a stormy session. Sitting on metal folding chairs in a hot, crowded, grade-school auditorium, their faces tense and strained with anger, the Buffalo Creek miners had risen one by one to blast a coal company statement which described the killer flood as "an Act of God"— an act for which, presumably, the company would not be liable. Most of these miners were inarticulate, sputtering with rage as they talked about the dam's construction, its lack of inspection, the absence of any warning system, and the previous breaks with should have signaled the oncoming disaster. They were inarticulate—but not helpless. The survivors committee was happy to hold up its meeting for half an hour so that the national television networks could get their cameras going. The organizers of the meeting had come prepared with an armload of mimeographed statements from witnesses, all of which were intended to prove that The Pittston Company was liable for the disaster. And, as the long meeting continued, the miners discussed questions like how to obtain the best possible legal counsel; how to make the most effective use of the media; how to bring the maximum amount of pressure on the politicians. Everyone agreed that it was finally time for the "little man" to stand up and fight.

But the most emotional moment of the long day came when Wayne Brady Hatfield stood at the microphone and told them about the death of his family. The miners squirmed in their seats as he described, detail by detail, the hours of horror which had changed his life. "I've got nothing left to lose now," Hatfield said. "I've already lost my wife and daughter, the two I loved so well. . . . I lost the lovingest woman who ever lived on this earth. I know it. No better woman ever lived on this earth.

"And I want to know why."

This book seeks to answer his question.

ACKNOWLEDGMENTS

MANY people helped in the preparation of this book. Thanks first to the residents of Buffalo Creek, West Virginia, for their patience during lengthy interviews. Particular thanks to Leroy Lambert, his wife, Easter, and sons, Roger Dale and Anthony, for their cheerful hospitality and their lessons on the history of Buffalo Creek.

Thanks also to Emery Jeffreys of the Logan Banner for showing me around Logan County; to the Peach Creek, West Virginia, YMCA, for the best $2-a-night room I ever stayed in; and, particularly, thanks to my friend Kathleen Lilly, whose patience with this manuscript—and the author's struggles to produce it—approached the heroic.

Warnings

CHAPTER 1

3 AM - A Call for Help

THE FIRST WARNING came in darkness, in the predawn hours of Saturday, February 26, 1972, with a cold, late-winter rain slanting across the mountains of southern West Virginia and a telephone ringing in the offices of the Logan County Sheriff.

Larry Spriggs heard the telephone ring, and frowned.

Burly and gravel-voiced, Deputy Spriggs was the night-shift jailer at the courthouse building on Logan's downtown Stratton Street. Since midnight, he had been sitting quietly in his office near the jail, smoking one Pall Mall cigarette after another and listening to occasional conversations on his police radio. So far, his shift had been almost routine, with only a few calls—a mud-stalled car here, a speeding driver there—to worry about.

Almost routine. But not quite. Because Logan County was under another of those "flash flood" warnings tonight, and although this was hardly unusual for West Virginia in late winter, it meant that Spriggs would have to remain especially alert. The flash flood watch, posted by the U.S. Weather Bureau in Huntington at 9 P.M. on Friday, had advised residents of a dozen West Virginia counties that, after two days of rain, serious flooding might be only a few hours away. Spriggs remembered the thunder and lightning of a few hours before (thunderstorms were supposed to be rare in West Virginia during February), and he remembered the way the heavy rain had lashed his windshield as he drove to work.

Of course, dangerous flash floods were nothing new in Logan County, where more than 46,000 people live mostly along unpredictable rivers and creeks which make up the Guyandotte River watershed. Perched high on the western flank of the Appalachian Mountains, about fifty miles south of Charleston, Logan County is 456 square miles of narrow, twisting valleys, jagged ridgelines, and racing creeks where the briefest rain squall can send floodwaters shooting through a valley like bullets through the barrel of a gun. Each year the spring rains bring their usual quota of flooded basements, washed-out roads, and rock slides, and every five years or so the county suffers through more severe flooding which leaves devastated homes and occasional drowning victims in its wake.

After years of dealing with such emergencies, the sheriff's office had worked out a regular routine: deputies patrolled the areas where flooding was most likely to occur; the office had put together contingency plans designed to warn residents of impending danger, and, if necessary, to call out the National Guard. In addition, Spriggs had received a special set of instructions at the start of his midnight shift: if the flood situation became dangerous, he was to call Logan County Sheriff Ralph Grimmett at home. Grimmett would then assess the situation, and personally decide on the best way to handle it. Now, as he lifted the telephone receiver to his ear at three o'clock on this wet Saturday morning, Spriggs thought himself well prepared for whatever the weather could do.

But he was not prepared for this telephone call. The caller identified herself as Mrs. Maxine Adkins, a thirty-nine-year-old practical nurse who lived in a tiny village called Three Forks, about twenty-five miles southeast of Logan, and she said she had a request to make of the sheriff's department.

She wanted Spriggs to call out the National Guard.

The stocky jailer sat straight up. The National Guard? For what? So far, there had been no reports of any major flooding at all, and even if the Guyandotte were to come over its banks, the floodwaters could never reach Three Forks, more than twenty miles away. What was the nurse talking about?

Mrs. Adkins hurriedly explained her problem: Three Forks, a cluster of twenty-three houses located at the top of nearby Buffalo Creek Valley (or Buffalo Creek "Hollow," as the residents refer to it), sat alone and exposed in the very middle of the narrow valley floor—only a few hundred yards below the base of a gigantic, earthen dam. The dam was the problem. Composed entirely of crumbly, unstable coal wastes, it stood sixty feet high and 550 feet long at the center. Behind the dam, millions of gallons of rainwater had backed up for more than half a mile. A terrible disaster was about to strike Three Forks—and the rest of Buffalo Creek Hollow as well—Mrs. Adkins said, because she had reliable information that the Three Forks dam was about to break.

"They say the water is rising six to eight inches an hour," the nurse told the jailer, "and that it's going to break loose before morning. If that water ever comes out of there, it'll take this whole hollow with it. You've got to evacuate the hollow before it's too late!"

Spriggs blinked. More than 5,000 people lived in the sixteen little towns along twisting Buffalo Creek. He wasn't about to rouse them out of bed on the say-so of a frightened nurse. Moving cautiously, the deputy wanted to know where Mrs. Adkins had gotten her information.

Breathlessly, the nurse told Spriggs that some of the older, more experienced miners who lived along the hollow had been watching the dam during the night. The water was rising fast, the miners said, and it looked to them like the big dam was in trouble. Several families (including Mrs. Adkins's own), frightened by these reports, had already evacuated Three Forks, the first town in the path of the water. Since midnight, they had been gathered in the red-brick Lorado schoolhouse, about three miles downstream.

Spriggs was attentive, but unconvinced, until Mrs. Adkins suddenly said, "A lot of the men have been talking to Jack Kent, and he says the dam is in bad shape. He's been telling people that it may go before daybreak—and that they better be prepared to get to higher ground."

Who was Jack Kent? A top official at the Buffalo Mining Company, which owned and operated the dam. Kent, the overall superintendent of strip mining for Buffalo, was the No. 2 man at the scene, and if he was that concerned, then the situation might really be critical. Now Spriggs decided to move. Thanking Mrs. Adkins, he promised her that the sheriff's department would check out the reports and then, if necessary, do its best to evacuate the 5,000 residents of the hollow. The deputy hung up. A moment later, following his instructions to the letter, he dialed Sheriff Ralph Grimmett at home.

Suddenly, the flood situation seemed dangerous.

The rain fell in darkness, and across the mountains of southern West Virginia, the people slept.

They slept at Skygusty and Cucumber, in rural McDowell County, where the deep rain drummed sudden rhythms on their wood-slat shacks, on their splintered barns, on the rusty bones of their junked-out cars. They slept at Dingess, and Parsley, and Red Jacket, in sparsely populated Mingo County on the Kentucky line while the late-winter rainstorm drifted down the Tug Fork River, glistened along the black asphalt roads.

And in the sixteen Logan County towns strung like a necklace down the middle of Buffalo Creek Hollow, they also slept, 5,000 human beings, as the rain came funneling down and the long night advanced.

They were coal miners, mostly, or related to coal miners, people with names like Hatfield, Osborne, Adkins, Trent: direct descendants of the rugged, Scottish-Irish settlers who crossed the Appalachians to build a new, pioneer life for themselves during the eighteenth century. Most of the towns they lived in—Lorado, Lundale, Amherstdale, Braeholm, and a dozen more—had been put together by the coal companies during the early 1900s, when coal in West Virginia was just getting its start. Many of these Buffalo Creek residents lived in monotonous rows of company-built homes, the kind of sagging, slat-porched houses which can be seen today on almost any hillside or bottom in the Appalachian coalfields.

But they were not poor. By now, coal had recovered almost completely from the terrible slump of the 1950s— when fierce competition from oil and gas in the energy market had forced total automation of the coalfields and cost tens of thousands of miners their jobs. The crisis had been painful: in West Virginia, for example, as the mines went from old-fashioned pick-and-shovel to new, automated machinery, employment in the coalfields fell from 125,000 during the peak year of 1948 (when 169 million tons of coal were produced) to 43,000 in 1961 (when only 111 million tons were produced). Displaced by gigantic mining machines, thousands of West Virginia coal miners —like their brothers all across Appalachia—had left their homeland to find jobs in the big, industrial cities of the midwest: Chicago, Detroit, Cincinnati.

But for the men who remained behind, the 49,000 miners who now worked in the state's 850 underground and 300 surface mines, the living had improved enormously. For one thing, the new machinery eliminated much of the back-breaking labor which had been the

miner's lot for two centuries. For another thing, coal
miners were making good wages these days—around $40
a day for the average digger, under the newly negotiated
United Mine Workers labor contract. And the future
looked even better: by the third year of the current contract, miners with special skills (electricians, for example)
could look forward to making as much as $50 a day, be-
fore overtime.

The future looked increasingly bright for the coal operators as well. The gloomy predictions of a decade before
—that most of the big utility companies would convert
from coal to nuclear energy by the 1970s—had proved
false. By 1972, as the American electric power shortage
intensified, fuel-hungry utilities were demanding ever-
increasing quantities of soft, steam-producing coal for
their giant turbines. In addition, foreign markets (particularly Japan, where booming steel and shipbuilding indus-
tries devoured coal as fast as it could be delivered) were
expanding at a spectacular rate. After enjoying, during
1970, its best single year since 1947, the coal industry was
talking excitedly about a future that looked increasingly
bullish.

Along Buffalo Creek these days, they were enjoying the
boom. Most of the men who lived in the little towns along
the creek worked—as their fathers had before them—at
the deep underground mines located in the mountains
high above the hollow. They worked as coal loaders, cutting machine operators, roof bolters, buggy drivers: doing
the thousand-and-one jobs that go with taking coal out
of the earth and sending it away on long black trains to
markets around the world. Each day, they drove the narrow, winding mountain roads to deep mines run by Buffalo Mining, Amherst Coal, Island Creek, and half a dozen
other coal operators as well.

Already enjoying a substantial wage, many of the

miners were cashing in on additional hours of overtime, as the coal companies struggled to meet the increasing demand for their product. By 1972, Buffalo Creek was anything but a depressed area: sleek, new trailers were going in next to the rows of company-built homes which had lined the hollow for decades; new housing construction was underway at several locations; many of the simple, wood-frame homes purchased by hollow residents from the coal companies over the years had been remodeled and brightly painted. In addition, many miners now owned a second car, and the average home now featured at least one color television set.

There was a price to be paid for this relative affluence, of course, a terrible price. But then, the miners had always known that theirs was a dangerous, dirty occupation. It was an occupation in which more than 100,000 American men had been killed since 1900—in which more than 1.5 million men had suffered serious injury since 1930 alone. With by far the worst safety record in American industry, coal was a business in which workers almost expected to lose a few fingers or toes during a career—if they were lucky enough to escape death in a slate fall or underground explosion. The federal government passed safety laws; the state governments passed safety laws; the deaths and injuries continued with monotonous regularity. Over the years, in fact, the constant threat of death or injury at the mine had simply become an accepted fact, a way of life, all across the Appalachian coalfields.

Moreover, increasing evidence had been assembled during recent years to show that the most dangerous threat to the coal miner was actually an invisible one: the dreaded miners' pneumoconiosis, or "black lung." Research now showed that almost every man who spent more than ten years underground could expect to contract the disease to one degree or another. Black lung meant breathlessness and lack of energy for the less severely afflicted; for the bad cases, it meant forced retirement at an early age, followed by gradual strangulation as they choked on the black coal dust which had clogged their lungs; and then finally, an early, agonizing death.

These were routine dangers: the possibility of death or injury in a sudden mine fall, the probability of sooner or later catching black lung, and the miners of Appalachia had long ago grown accustomed to living with them. But there was another, less obvious, but even more ominous hazard which the people of Buffalo Creek faced daily: the gigantic coal refuse pile which sat, smoking and smoldering, at the top of their hollow. After twenty-five years of steady dumping by three different coal companies, the coal waste (or "gob," as the residents refer to it) had by now formed an immense, 200-foot-high, 1,500-foot-long barrier. At the upper end of the pile, the gob had simply been dumped—and then compacted with bulldozers—to form a retaining dam which now impounded more than 130 million gallons of water.

Actually, the situation was not at all unusual for the mountains of Appalachia, where enormous, often-smoldering gob piles—the waste products from coal processing— can today be seen on almost every hillside. By 1972, in fact, the big coal companies had dumped more than 900 such piles of black, sooty refuse across coalfields from Pennsylvania to Alabama. The Buffalo Creek dump was different only in that it was larger than most and because of the immense amount of water which had backed up behind it.

The dumping process was simple enough. The coal companies mined their valuable black mineral from both deep and strip mines which were usually located high in the mountains. The raw coal went by conveyor belt or railroad car to the company processing plant, or "tipple," where it was separated from the rock around it and washed. The finished product was then shipped out to markets around the country in long coal trains which chugged night and day through the steep mountains.

But the waste materials—chunks of rock, or slate, picked up during the mining, slabs of impure, discarded coal, and tons of coal dust—were left behind, to be dumped in the nearest open space. In the rugged mountain country of West Virginia, open space is at a premium, and so the companies usually wound up dumping their refuse in the closest available hollow.

At Buffalo Creek, the process had been going on since 1947, when the now-defunct Lorado Coal Company (a local concern) built a tipple to wash coal coming from its newly opened Mine No. 5, located in the mountains above the hollow. At first, the company dumped its gob on a hillside near the tipple. After a few years, however, this space was exhausted and Lorado Coal began shifting the gob to the floor of Middle Fork—one of three small, stream-bearing hollows which funnel into Buffalo Creek Valley just above the village of Three Forks (or Saunders, as it is described on maps).

As the years passed, the gob pile inched farther and farther up Middle Fork Hollow. Gradually, as the smoking dump (coal waste piles often ignite spontaneously and then burn deep inside for years, glowing like charcoal) invaded the narrow hollow, residents were forced to sell their homes and move out. In the end, Middle Fork was deserted, occupied solely by the monster coal dump, and the dam which stood at its head.

Meanwhile, Lorado Coal had gone out of business, and the entire operation had been purchased, in 1964, by another local company, Buffalo Mining. With the purchase of Lorado's holdings, Buffalo inherited two problems: first, great quantities of water (the result of impounding both

Middle Fork Creek and years of collected rainwater) had accumulated behind the dump, posing a hazard to families who lived below it.

Second, new statewide anti-pollution laws had been passed which prohibited the company from draining its black, sludge-filled water into once-clear Buffalo Creek. The company solved both problems at once by creating, during the next several years, a system of retaining dams and settling ponds on the valley floor above the big dump. The ponds allowed waste materials to settle out of the dirty water—some of which was then pumped back to the tipple to be used again, and some of which filtered through the dams and the dump below them to emerge, purified again, at Buffalo Creek. The first of these dams, built to a height of about twenty feet, was put together by simply bulldozing the gob into the proper configuration.

The system worked well for a while, but after a few years the once-porous dam had a tendency to silt up, preventing seepage and thus allowing the black water to overflow its impoundment and again pollute the creek. To solve this problem, Buffalo Mining in 1966 built a second dam, also about 20 feet high, 600 feet farther up the hollow. The second pond fulfilled its function for a few years, and then also proved inadequate.

Finally, Buffalo Mining decided to build a system which would be failureproof. Moving another 600 feet up the hollow, the company in 1969 began dumping tons and tons of gob to form an enormous dam which would hold all the water the company could ever need. By February 1972, dam No. 3 extended 550 feet across the valley at its center. From front to back, it was more than 400 feet thick in most places. And along its crest, the new dam stood fifty to sixty feet high above the valley floor. Behind the dam, about half a mile of water had collected to a depth of perhaps thirty feet. Beneath the water, another twenty feet of sludge had settled to the valley floor— which meant that during normal times, the water level remained about 10 feet below the crest of the dam.

While the Three Forks dam was being constructed, the property on which it stood, and the mines around it, changed hands again. This time the buyer was a giant in the industry, a New York-based conglomerate which did $500 million worth of business each year—in coal, oil, trucking, warehousing, and (appropriately enough for a company with more than $400 million in assets) armored car protection services. Growing furiously, and always on the lookout for coal reserves which could help it meet its accelerating contract requirements, The Pittston Company, the nation's fourth-largest coal producer, purchased the Buffalo Mining Company in June of 1970.

The day-to-day operations of the Buffalo Mining Company were hardly affected by the purchase, however, since The Pittston Company system called for giving its many coal subsidiaries as much independence as possible. Buffalo Mining acquired new leadership, of course, appointed by Pittston's New York headquarters, and the little subsidiary had its major decisions made for it by the corporate bosses on Park Avenue. But in the eight coal mines which Buffalo operated around the hollow, the work of mining and shipping coal went on pretty much as it always had. And the coal waste dam at Three Forks, about half completed when Pittston took over the company in June of 1970, was quickly finished.

Meanwhile, nobody at Buffalo Mining or The Pittston Company seemed to be concerned—or even aware—that the giant dam had been put together without the slightest attention to engineering safety. Built on a base of twenty feet of gooey, unstable sludge (the deposits from settling pond No. 2), haphazardly constructed out of every kind of mine refuse, from slate chunks to old blasting wire to rotting timbers, the hastily built barrier would be expected to contain up to twenty million cubic feet—or more than 120 million gallons—of water in Middle Fork. In the face of this requirement, as government engineers would later point out, Buffalo Mining's dam-building technique was absurdly unscientific: the gob was simply dumped on the dam's surface by trucks and then compacted with bulldozers.

Without any written engineering plans (none was ever prepared, as it turned out), and with no scientific way of assessing the stresses which the dam would have to face, the company officials really had no idea at all of whether or not it would hold. (Later, in fact, after the disaster, government engineers and geologists would conclude that the barrier had been doomed from the start and that, under the pressures it faced, the failure of February 26 was inevitable.)

If there seemed to be little concern at Buffalo or Pittston headquarters about the dam's safety, there was apparently less concern for the fact that its construction and maintenance were in clear violation of at least two laws. First, under a West Virginia Public Service Commission regulation, anyone who built a ten-foot barrier which obstructed a water course was required to obtain a permit. Pittston sought no permit in building its sixty-foot dam, and the PSC later confirmed that this was a violation of the statute.

Second, a U.S. Bureau of Mines regulation, authorized by the Federal Coal Mine Health and Safety Act of 1969, required coal companies which maintained hazardous water impoundments to inspect them at least once a week. Written reports of the inspections were supposed to be filed with the local Bureau of Mines office. No inspection reports on the Three Forks dam were ever filed with the

Bureau of Mines, and the Bureau later cited the company for breaking this law.

The potential for disaster was further heightened by the fact that Pittston had devised no warning system of any kind—no plan to alert the valley below, or the local authorities, in case of an impending break in the dam. As later evidence indicated, company officials simply refused to believe—even as the ominous signs of approaching catastrophe multiplied—that their unengineered dam could give way.

And so the 5,000 people of Buffalo Creek Hollow slept on through the rainy night. At Lorado and Lundale and Stowe Bottom; at Crites and Pardee and Accoville; in double rows of houses laid out between the state highway (W. Va. 16) and the railroad tracks which carried the coal away, they slept and passed the long winter night. And in the reservoir at the top of the hollow, behind an unstable, illegal dam which never should have been built, the rain came slanting down the wind and the water rose inch by inch.

A few residents remained awake, however.

At the red-brick grade school in Lorado, about three miles below the threatened dam, several families from Three Forks had been gathered since midnight. Having lived under the shadow of the Buffalo Mining dams for many years, they knew better than anyone else the danger they faced. They knew that for years the lake behind the Buffalo Mining impoundment had been deep enough to run motorboats on—and that, during periods of heavy rain, the water often rose close to the crest of the dam. Now, as they sat drinking coffee and nibbling cookies at three o'clock in the morning—Roy Chandler and his family, Denny Gibson and his wife and four kids, Preacher

Herbert Bailey, and half a dozen others—they wondered if the disaster they had feared for so long was about to happen.

Pearl Woodrum, a graying, elderly widow who lived with her daughter Velma and her dog, Critter, at Three Forks, had a special reason to be afraid. Like the other residents of Three Forks, Pearl had lived through an unnerving, partial break in the giant impoundment (at dam No. 2, in early 1967). The water had come roaring out of Middle Fork then, ripping out the highway, a section of railroad track, and several basements. Like the other residents, Pearl had been terrified by the experience, and asked herself what would happen if all the water in Middle Fork ever escaped at once.

Unlike the others, however, Pearl Woodrum decided to do something about the problem. She wrote a lengthy, passionate letter to then Governor Hulett Smith, asking him to help in eliminating the hazard. "We are all afraid we will be washed away and drowned," Pearl wrote, explaining that every time it rained hard, people in Three Forks sat up all night worrying about the dam. "Please, for God's sake, have the dump and water destroyed. . . . Our lives are in danger."

Governor Smith answered the letter. He told Mrs. Woodrum that he was taking action—the letter had been passed on to the state's Department of Natural Resources, which would take a hard look at the situation. A few days later, Pearl was visited by a man from the DNR: he told her he had just inspected the Buffalo Mining complex and she was right, the dam was certainly dangerous. The DNR man promised her the danger would be immediately eliminated, and left. After that, Pearl never heard from the state of West Virginia again.

What happened next is not entirely clear. The DNR

apparently asked Buffalo Mining to make some minor
alterations—nothing which materially changed the hazardous situation—and the matter was dropped. After the
disaster, Mrs. Woodrum's 1968 letter surfaced again, and
it became clear that the problem had simply been shuffled
from department to department in the state government.
State officials insisted they had forwarded the file to the
Logan County Prosecutor's office, with a suggestion that
Buffalo Mining be prosecuted for breaking the state law
which requires dam-builders to obtain a permit.

But the prosecutor's office claims that the state sent it
the file "just for the record," while assuring the office that
the problem had been cleared up by the DNR. "You could
tell just by looking at the correspondence," said an assistant Logan County prosecutor after the disaster, "that
all they had done was pass the buck back and forth. They
weren't interested in cracking down on Buffalo Mining—
not by a long shot."

There were other, equally ominous forewarnings during
these years of the disaster which might one day strike
Buffalo Creek. In 1966, for example, an enormous coal
waste dump collapsed near the mining town of Aberfan,
in Wales, burying 147 people—most of them children
from a local school—under tons of water and gooey
sludge. Alarmed, the U.S. Department of the Interior immediately ordered a safety study of similar refuse dumps
in the Appalachian coalfields. Performed by the U.S. Geological Survey, the study found that thirty of the thirty-
eight structures it examined in West Virginia were unstable, and four of them were critically dangerous. The
Three Forks dump was one of those inspected: U.S. Geologist William Davies found it basically stable, but noted
in his report that the dam in use at the time (dam No. 2,
1967) "could be overtopped and breached." Davies went

on to warn that "flood and debris would damage church and two or three houses downstream, cover road and wash out railroad," if the dam ever did give way.

The Department of Interior study showed clearly that coal waste dams were a dangerous hazard throughout the Appalachian coalfields. Dutifully, the Department mailed out copies of its report to the offending coal companies, to all of the applicable government agencies, and to various state and federal office-holders as well. The report made the newspapers for a few days, and then was quietly forgotten.

It wasn't long after Aberfan, however, before a collapsing coal dump again made newspaper headlines in West Virginia. This time the collapse took place much closer to home—at a place called Proctor Hollow, right on Buffalo Creek, in fact—where, in the summer of 1967, a slide from an Island Creek Coal Company gob pile thundered through an entire neighborhood, filling basements with mud and sweeping away several automobiles. U.S. Congressman Ken Hechler, who represented southern West Virginia, called a press conference to blast the coal companies for creating such hazards in his district. Hechler summoned several state officials to the scene, and told them he was "horrified" by what had happened. "These people are living under the gun of threatened annihilation," Hechler charged, while demanding that the coal companies take immediate steps to correct the problem.

But the months passed, the newspapers turned to something else, and soon the situation reverted to the way it had always been: with the people who lived in places like Buffalo Creek Hollow well aware of the threat to their lives and property, and the state and federal governments either insufficiently concerned, or simply unwilling, to take the decisive hard-nosed steps needed to solve the problem.

Of course governmental foot-dragging on questions of coal mine safety was nothing new. In spite of the spectacular mine disasters which periodically caught the nation's eye, and in spite of the coal industry's terrible safety record, mine safety laws had been agonizingly slow to evolve in the United States. The first federal safety bill, which created the U.S. Bureau of Mines under the Department of the Interior in 1910, had contained no provisions for enforcement at all. The accidents continued with regularity (16,834 coal miners have died in West Virginia mines alone, for example, since the first law was passed), until a 1951 disaster which killed 119 men at West Frankfort, Illinois, finally propelled Congress into action.

The new law was broader, but still lacked penalties for violations. And the disasters kept making headlines. Finally, on November 20, 1968, an explosion deep inside the Consolidation Coal Company's No. 9 mine at Farmington, West Virginia, killed seventy-eight miners and focused the entire country's attention on the problem of coal mine safety.

Farmington became a nine-day-long nightmare—covered from beginning to end on nationwide television—in which explosion after explosion rocked Consol's 600-foot-deep shaft mine, preventing rescue workers from reaching the doomed men inside and gradually erasing the hopes of the wives and children who waited pathetically outside the portal through which they would never return. After the nine days of horror at Farmington, Congress moved quickly to pass the nation's first strong safety bill in 1969. The new regulations bristled with teeth, but the undermanned U.S. Bureau of Mines often had trouble enforcing them. The fatality rate in U.S. coal mines was declining rapidly by 1972, but no one doubted that coal mining was still the most dangerous job in the world.

If the federal safety laws were weak, the state laws could best be described as lifeless. In states like West Virginia, where coal interests have always maintained tight control over the legislatures, mine safety laws were usually so watered-down as to be almost useless. In addition, West Virginia and most of the other Appalachian states were notorious for failing to enforce the few laws they did pass. Typically, the miners say, an inspector would arrive after the coal operator had made thorough —and temporary—preparations in advance. After checking off a few, routine violations—for lack of safety helmets and exposed wiring, maybe—he would depart, not to be seen again for as long as a year. In effect, the miners knew they were on their own, protecting themselves from accident as best they could.

The situation was made even more hopeless by the fact that the miners' supposedly biggest ally—the once-powerful, once-prestigious United Mine Workers labor union— had become, during the 1950s and 1960s, one of the most corrupt labor organizations in the United States. During those years, the same union which had delivered the miners from virtual slavery in the early twentieth century demonstrated again and again that it had more in common with the big coal operators than with the men in the mines. The UMW had, in fact, entered into a virtual partnership with the owners, allowing them to mechanize their mines at the expense of thousands of jobs and even loaning them millions through a UMW-controlled bank in Washington. Gradually, the money-hungry leadership had trimmed thousands of once-eligible disabled miners and miners' widows from its pension fund. By 1972, the union was controlled by convicted criminals who rigged elections, misspent enormous sums of money, and used strong-arm tactics (perhaps even murder) to stifle opposition. By this time, the *UMW Journal*, once regarded as the miners' very Bible, had become a pompous propaganda sheet which lapped praise on the union's leaders, headed by the notorious, hard-nosed Tony Boyle. The union had years ago ended its vigorous battles for improved job benefits and better safety conditions, and it was not about to begin insisting that the coal companies go to the heavy expense of removing hazards like the Buffalo Creek dam.

The coal company executives, the government officials and the union leaders seemed to care little about the deadly threat which faced the residents of Buffalo Creek Hollow on this rainy Saturday morning in late February. But the miners who lived in the sixteen communities along the hollow were well aware of it. Indeed, many of them had been running from the big dam at Three Forks for years. During periods of heavy rain, families would often pass the word from Three Forks, the first town in the path of the water: They say the dam's about to go, better get ready to run. Some families remembered spending entire nights on hillsides, sleeping under tents and makeshift lean-tos, in order to assure their safety. Once in a while, even, practical jokers would race up and down the narrow road that ran the length of the hollow, honking their horns and shouting that the dam had broken.

Remembering the past, then, some families had slept fitfully, or not at all, on Friday night. Again and again, these residents had braved the rain to go outside and take another look at the creek. It had risen only a little during the night, which was comforting, even though the creek could tell them nothing about the stability of the dam at the top of the hollow. As they checked and rechecked the depth of the stream near their homes, these vigilant ones had no way of knowing that, miles above them at the dam site, at least one company official already considered the situation so dangerous that he was preparing emergency plans to prevent a disaster.

They had no way of knowing that the worried official would soon begin assembling a crew of workmen armed with bulldozers, endloaders, and other heavy construction equipment, in a last-ditch effort to save the dam. And they could not have guessed that, increasingly frightened by the possibility of an imminent catastrophe, Buffalo Mining Company superintendent Jack Kent had already warned several of his neighbors to prepare to run.

The rain continued, sloppy and insistent, gusting off the mountains above Buffalo Creek Hollow as though it would never stop. At the brightly lit Lorado schoolhouse the frightened families sat patiently, waiting for the safety of daylight. And twelve miles to the south, in the little town of Man—where Buffalo Creek joins the Guyandotte River—Logan County Sheriff Ralph Grimmett was trying to decide what to do.

Grimmett, a wiry, soft-spoken man, a successful auto parts and heavy equipment dealer who had recently been appointed sheriff, could not have been happy to hear jailer Spriggs's report of trouble at the dam. Instinctively, the sheriff sensed the one thing he dreaded most: political problems. Facing a tough primary election fight the following May (and then the general election in November, he hoped), Grimmett realized unhappily that whatever course of action he decided on might cost him votes.

Part of the problem was simply that this was Logan County. Described by outsiders and local residents alike as "the most politically conscious county in the United States," Logan has a strange, tortured history of political corruption, vicious factionalism, and bloody violence in which only one factor seems to remain constant: coal money usually controls elections.

Coal's kingdom in Logan County had been established soon after 1904, when the Chesapeake and Ohio Railroad completed its Charleston to Logan line, and thus opened

East Coast markets to the county's rich deposits of bituminous coal. (Logan ranks second in coal reserves among West Virginia's forty-six coal-producing counties, with more than 3.9 billion mineable tons of the mineral remaining in its thick mountain seams. The county ranked fourth in the state in coal production during 1971, shipping out more than 10.1 million tons.)

By 1910 more than fifty coal companies were shipping millions of tons of the valuable mineral to markets all over the East, and the countryside was dotted with dozens of booming coal towns. The coal towns were virtual slave camps, where thousands lived in primitive shacks without plumbing, bought their groceries with company-issued scrip at company-owned stores, and worked twelve to fourteen hours a day in gloomy, death-trap mines where accidents killed or injured them on every shift.

Oldtimers who had grown up around the mines of West Virginia could still remember terrible years like 1908, when 625 miners died on the job in their state; or 1914, when 541 had perished; or 1925, when another 686 were killed in the mines. West Virginia held the record for the biggest coal mine disaster in U.S. history—a 1907 explosion at Monongah in the northern part of the state which killed 361. And for every death there were perhaps a dozen injuries: broken backs, severed legs, crushed hands. It had been an era in which ambulances often sat permanently parked at the mine portals; in which undertakers who did business in mining areas could count on a steady flow of customers.

During these early years, the coal operators controlled every phase of the miners' lives, and they soon controlled the local politics as well. By 1915 they had succeeded in bringing to power one of the most notorious characters in West Virginia's history, Sheriff Don Chafin, the "King of Logan County." Operating from his own private arsenal in the county courthouse, the ruthless Chafin gradually built an army of gun thugs and criminal misfits whose main task was to enforce the will of the coal owners. Short, stocky, and dark-complexioned, a hard-nosed bully with an enormous appetite for power, Chafin would hire as many as 400 deputy sheriffs at a time. Wearing pistols on their hips and carrying heavy clubs, the deputies were dispatched to the mines with instructions to sniff out and eliminate union organizers.

Other deputies watched the Logan train station. Strangers who alighted from arriving trains found themselves confronted by two or more thugs who demanded to know the visitor's business. If a newcomer refused to talk—or if the deputies didn't like his story—he found himself still on the train when it pulled out of Logan station. Those who resisted, like the visiting state official from Charleston who thought himself immune from Chafin's power and wound up lying in the road with multiple injuries, were beaten and sometimes killed.

Old-time miners in Logan County remember that, during those early days, gun thugs stood outside every mine portal watching relentlessly for union organizers or even sympathizers. Company spies infiltrated the ranks of the workers, and were prepared to inform on their fellow miners for the slightest infraction of the anti-union rules. "Everybody was scared," says seventy-two-year-old Willard Adkins, a white-whiskered, hollow-cheeked miner who worked in the Logan mines during the Chafin era. "You couldn't say nothing to nobody. If them thugs saw two or three men talking together, why, they'd run up and club 'em apart. Everybody hated the thugs, but there wasn't nothing you could do about it except kill 'em—and I guess a lot of 'em got killed, too!"

The coal companies paid Chafin a fee for each mine guard he provided, and by keeping a percentage for himself, the sheriff gradually built a fortune. By 1920 the owners were willing to pay a premium for their mine guards because, as the unions gained strength in the coalfields of northern West Virginia, it was getting harder and harder to keep them out of Logan County. In 1921, increasingly frustrated by their inability to organize the booming Logan coalfields, and vowing that they were going to end "Czar" Chafin's rule forever, 4,000 miners gathered outside Charleston for what was to be one of the strangest, and most violent, labor-management confrontations in U.S. history.

Led by union organizer Bill Blizzard and a fiery, ninety-one-year-old radical grandmother named Mary "Mother" Jones, the angry miners rounded up every hunting rifle they could find. While Mother Jones harangued them to "Go out and kill every goddam gun thug in West Virginia!" they drew up military-style plans for an invasion—and hopeful "liberation"—of Logan County. And then, marching in ranks and carrying one old rickety machine gun, they trooped south along the Little Coal River toward Logan. Meanwhile, Chafin had put together a small army of several thousand men, made up of gun thugs, hired criminals, jail inmates, and state police (the state police fought with Chafin because, as sheriff of the county, he represented the "duly elected authority").

Chafin's army took up positions along a ten-mile ridgeline at the northern end of the county, and the stage was set for what later became known as the "Battle of Blair Mountain." As the miners made futile attempts to scale the ridgeline, a Chafin-hired private pilot dropped bombs on them. The bombs all missed, but the rifles didn't and there were numerous casualties before U.S. troops arrived to restore order. Chafin had won the battle for Blair Mountain, and it was another fifteen years before Roosevelt's National Industrial Recovery Act (1936) broke the back of union resistance and the Logan coalfields were finally organized. Chafin retired soon thereafter—with an estate valued at $300,000—and lived through twenty years of plush retirement before dying of old age in a Huntington hotel penthouse suite.

By now, political corruption and bloody violence had become a way of life in Logan, and the county was gaining a statewide reputation as a haven for criminals of one kind or another. In one celebrated case, a zealous Logan police chief who unwisely attempted to shut down a widespread slot-machine gambling operation was murdered in a local restaurant by a skinny, professional killer named Enoch Scaggs, who shot the chief in the back five times while the latter stood paying for a cigar.

The killing stirred a blizzard of protests—as well as snickers—across West Virginia, and the state attorney general decided that in order to obtain justice, he would have to prosecute the case himself. Both the county prosecutor and the sheriff, who were thought to have been linked to the murder, if not directly behind it, were barred from the proceedings. In a courtroom flanked by several state police, Scaggs was convicted, but not before the attorney general discovered that witnesses were sneaking through an open courtroom window to be coached on their testimony in an adjoining room by the corrupt prosecutor and his men.

As the years passed, and Logan County's reputation worsened, the residents began to consider themselves celebrities of a sort, enjoying their notoriety in the same way they had enjoyed it during the late nineteenth century when the Hatfield-McCoy feud raged across the southern part of the county and brought the entire nation's attention to them. Politicians in Logan County changed sides so fast, joining and quitting opposite "factions" with such rapidity, and with so much venom directed at their opponents, that politics in the county began to be seen as a kind of sport: a complex, kaleidoscopic game in which nothing was certain except that the man who was up today would be down tomorrow—if only you were patient enough.

"I don't know how to explain it," says one local politician of the volatile, constantly smoldering political atmosphere in his county. "They just seem to love the game. You take some of these businessmen around here—they'll watch every dime they spend on their businesses, but then think nothing of blowing $50,000 on an election, even if it's hopeless from the start. And impeachments! God, every time you turn around they're trying to impeach somebody else. These people see impeachment as a regular part of the political process."

In a county long accustomed to rigged elections where money and whiskey flow in torrents and voters who have been dead for decades suddenly take life to appear in a voting booth, and in a political atmosphere so clouded that "corrupt" had become a regular party designation, like "Republican" or "Democrat" in other states, few people were very shocked when Logan County Sheriff Earl Tomblin, state senator Bernard Smith, and three other local officials were indicted by a federal jury for vote fraud in the spring of 1970. All five of them were later convicted and sentenced to prison terms.

It was in this murky, tangled political climate that Sheriff Ralph Grimmett—appointed to fill out Tomblin's term soon after the latter's indictment—would have to make his decision about Buffalo Creek. It would not be an easy decision. Waking up Buffalo Creek Hollow and alerting the National Guard could do nothing but cause problems for the Buffalo Mining Company, which would then have to answer a lot of questions about its dangerous dam and then possibly be forced to make expensive repairs. Grimmett knew that antagonizing Buffalo Mining and the other coal companies which maintained similar refuse dumps across the county could be political poison in an election year.

And the sheriff had other considerations which would make the decision difficult. One of his close friends and business associates was A. D. "Buster" Scaggs, the former owner of the Buffalo Mining Company, who had sold the entire operation to Pittston in 1970. Scaggs had put the Buffalo Mining Company together from the ground up, and he was the man primarily responsible for the big dam which now threatened the hollow. In addition, Scaggs was reported to have maintained close ties with Pittston—as well as holding a sizeable block of the New York conglomerate's stock.

If all of this was not enough to make Grimmett hesitate, it was also true that Buster Scaggs was running the political slate on which the sheriff hoped to be nominated for a second term in May and then elected in November. As a nominee for the county "court" (the three-man board of commissioners which runs Logan County), Scaggs headed a faction which most observers described as coal company-oriented: the established, conservative, machine organization in Logan County. Grimmett was running for Sheriff on Scaggs's ticket, in a campaign fueled by Scaggs's money, and he could hardly afford to call attention to the fact that the boss was responsible for the deadly hazard on Buffalo Creek.

On the other hand, however, the sheriff had to wonder what would happen if he ignored the calls for help and then the dam actually broke. Besides, the people gathered in the Lorado schoolhouse and their worried neighbors would be voting soon. It was important to let them know that when they called the sheriff, he came.

The first thing Grimmett did after hearing Spriggs's

description of the situation was call the Buffalo Mining Company to see what the bosses thought. The bosses thought everything was fine. When Grimmett reached Ben Tudor, a general superintendent and another top-ranking official at the big mining complex, Tudor told the sheriff that the reports of trouble at the dam were exaggerated. There was no need for alarm, Tudor insisted.

Grimmett then checked with other company officials (he will not identify them) who also assured him that the dam was stable and advised him to forget about the warning from Mrs. Adkins.

But for some reason, the sheriff wasn't convinced. "Some of these miners have been around a long time," he said later. "It's hard to fool 'em. When they talk, I listen."

Finally, almost two hours after the first warning call from Lorado, Sheriff Grimmett made his decision. He called jailer Spriggs back and told him to begin a full-scale alert of the sixteen towns along Buffalo Creek Hollow. The jailer was to call two sheriff's deputies, Otto Mutters and Max Doty, at home, and immediately dispatch them to the hollow, where their orders were to begin warning as many residents as they could find: The Three Forks dam is about to break, leave your homes at once and get to higher ground. In addition Spriggs was to inform the National Guard headquarters in Charleston that Buffalo Creek Hollow now faced a potential catastrophe, and ask the Guard for assistance with the evacuation.

Spriggs did his job. By 5:30 A.M. the deputies had been awakened, had dressed in their black-and-gray sheriff's uniforms, and were driving in separate patrol cars through the rainy darkness toward Buffalo Creek.

But the jailer was not as successful with the National Guard. The duty sergeant at the Guard's headquarters in Charleston gave Spriggs a telephone number—for his commanding officer, the one man who had the authority to send out the troops for emergency assistance. Spriggs dialed the number: nobody answered. He tried again. Still no answer. Irritated, the jailer called the sergeant back and explained the situation. "He told me he was sorry, but there was nothing he could do," Spriggs remembered later. "He said I'd just have to keep trying to get this major on the telephone. So I dialed the number every half hour or so, trying to get through. But I never did."

No clear explanation for the breakdown in National Guard communications has yet emerged. And the lapse would prove tragic: by the time the Guard finally learned of the potential disaster facing Logan County on this Saturday morning, the break in the dam was only moments away and it was too late to help.

Spriggs was irritated by his inability to reach the Guard, but at this point it didn't seem to matter very much. After all, deputies Mutters and Doty were already on their way to the hollow, and a full-scale warning was only minutes off. There was still time to prevent a major tragedy—if only the deputies did their job.

The bosses at the Buffalo Mining Company had assured Sheriff Grimmett that the Three Forks dam was in no danger, but at least one of them knew better. Jack Kent watched the rain spatter against the windshield of his green Ford pickup truck, and for the tenth time that night, tried to analyze the situation.

First, the water behind the huge impoundment at Middle Fork was rising faster than ever. By counting the notches on a wooden measuring stick which he had placed in the dam two days before, Kent could compute the rate of the rise fairly accurately: since midnight the water level had been climbing at about two inches an hour.

Second, as Kent had just noted on his most recent visit to the dam, around 4:30 A.M., the water was now lapping less than a foot from the top of the huge coal-waste barrier.

Third, the rain showed no signs of letting up.

The mathematics were compellingly clear, and ominous. If the rain continued, and if the rate of the water's rise remained constant, the Three Forks dam would be topped within a few hours. After that, no one could predict how fast the water would erode the soft, crumbly surface of the dam. Only one thing was certain: if the dam was ever topped, then the erosion would begin. And sooner or later, moving under tremendous pressure, the millions of gallons of water behind the dam would break loose and hurtle down the valley.

Driving down the hollow road toward his home in Lorado tucked safely away on a hillside, Jack Kent was getting tired. For two days now, it seemed like all he had been doing was making trips up to the impoundment at Middle Fork, checking the water level and trying to gauge the dam's stability. By Friday night, in fact, as the rain fell harder and the water climbed higher, an increasingly concerned Kent had begun returning to the dam every two hours. Somehow, without any explicit orders ever having been given, the 500-foot-long barrier seemed to have become his own, personal responsibility.

Kent had been a logical choice—the Buffalo Mining Company official most likely to wind up in charge of the dam. A quiet, balding, fifty-year-old man, strip mining superintendent Kent had spent twenty-four years of his professional life around installations like the one at Three Forks. Although he had studied only two years of mechanical engineering (at VPI, in 1939–40), Kent was widely respected among the miners for his common-sense, soft-spoken, practical approach to any problem which involved earth-moving and heavy equipment.

A Navy fighter pilot during World War II, Kent had

seen a lot of action while reaching the grade of Lieutenant Junior Grade. Since the war, Kent had worked for three different coal companies and had faced, and successfully handled, almost every situation which could arise in the tricky business of mining coal. He knew what he was doing.

And now, Jack Kent was worried. Traveling back and forth to the dam on this rainy Friday night, he had stopped several times in the village of Three Forks to visit with miners he knew (most of them also worked for the company). And the message he passed to each of them was the same: The situation looks dangerous. Better get your family out of Three Forks.

Denny Gibson was one of those who had visited the dam with Kent. Twice during the night, the strip superintendent had pulled his truck up next to Gibson's house in Three Forks and honked. On both occasions, Gibson had then accompanied Kent to the dam, where they had parked, climbed out of the truck, and walked out on the dam's surface.

On the first trip, around 12:30 A.M., Kent checked his measuring stick and told Gibson that the water had climbed one foot in the past six hours.

On the second trip, a few hours later, Kent noticed that the water was rising even faster. "He told me that if the water ever reached the top of the dam," Gibson remembered later "we were going to have to start evacuating the valley. He said, 'If all that water ever comes out of there, it's going to be real bad.'"

Gibson, who worked as a bulldozer operator for Kent's strip mine crew, had hours ago followed the advice of his boss by moving his wife, Wilma Jean, and his four kids out of Three Forks, three miles downstream to the brick Lorado schoolhouse, where he figured they would be safe no matter what the water did. Most of the other families in Three Forks had done likewise. Jack Kent usually knew what he was talking about.

Now Kent switched the engine off, climbed out of the pickup, and walked up the driveway toward his hillside home. They didn't have many options left now, with the water less than a foot from the top of the dam. Perhaps they could cut an emergency spillway in the dam, and that way relieve enough of the pressure to prevent a break. Maybe they could strengthen the barrier, by dumping additional tons of gob on its surface and packing it in. Maybe. But whatever action they took would have to begin soon. Further delay might soon prove disastrous.

His mind made up, Kent entered the darkened house located right above the Buffalo Mining office in Lorado, and picked up the telephone. It was time to call the big boss—Buffalo Mining vice president and general foreman Steve Dasovich—and let him know that the problems at the dam had become critical. Dasovich was the No. 1 man in the entire operation, and he would have the final say-so. Kent dialed, and a moment later, at Steve Dasovich's pleasant, tree-shaded home in South Man, the telephone was ringing.

CHAPTER 2

6:30 AM - Too Little, Too Late

Otto Mutters, a short, stumpy, cigar-smoking man who wears at all times the round, scowly face of an outraged bulldog, opened his eyes and stared at the ceiling.

What were they doing, waking him up so early? Only two and a half hours ago, Mutters had finished his late shift with the Logan County Sheriff's Department, and turned in for the night. It had been a routine shift, driving his gray, 1969 Ford sedan back and forth along the dark roads of Logan County's Triadelphia District—with only a couple of drunks to be locked up for the night, and a couple of speeders to chase. But the deputy was tired, and had dropped off to sleep gratefully, at 3 A.M.

Now he heard his fifteen-year-old daughter Cathy's voice once again summoning him out of a deep sleep, telling him to hurry up and answer the phone. Somebody was calling Deputy Mutters from the Logan County Jail, Cathy said, and it sounded important.

Groaning and stretching, Mutters padded across the cold floor to the telephone. The caller was Larry Spriggs, night jailer at the county courthouse, and he said he had special instructions from Sheriff Grimmett. The sheriff had just received word that the big dam near Three Forks was about to break. Mutters was to proceed immediately to Buffalo Creek Hollow, and begin warning residents of the danger. A second deputy would also be dispatched to Buffalo Creek to join Mutters. The two of them were instructed to start immediately a systematic, town-by-town alert.

Here we go again, Mutters thought. He didn't know much about the Three Forks dam—its size, or the amount of water behind it—but he did know that every time it rained hard on Buffalo Creek, the cries of "Wolf!" began again. At least once a year, it seemed, dire predictions of an imminent flood raced up and down the hollow—and nothing ever came of them.

But Mutters was a sheriff's deputy; he would do his job. Now he hung up the telephone and climbed into his uniform, pulling on a blue raincoat and matching rain hat for protection against the rotten weather outside. Five minutes later, he steped into the gray patrol car with the blue flasher on its roof and headed for Man.

Before beginning the long, wet ride up Buffalo Creek Hollow, Mutters decided to grab a quick cup of coffee. When he reached the three-block-long downtown section of Man, he pulled the car over next to the Man Smoke House, a small restaurant which he and his wife had been running for several years. The restaurant would be open, Mutters knew, because his wife always went in at 4 A.M. to ready the place for the day's customers.

As he waited for the coffee which would bring him back to life after only two hours of sleep, Mutters was surprised to see Steve Dasovich sitting at the counter beside him. Mutters knew who his fellow coffee-drinker was; he knew Dasovich's dam was the reason he had been so unexpectedly awakened. The two chatted amiably as they sipped from their cups. Dasovich asked Mutters what the deputy was doing up so early, and Otto told him about the problem at Three Forks.

Dasovich chuckled. He was in a friendly mood, Mutters thought. The general foreman explained that he had just been awakened himself by a report of trouble at the dam. But this was nothing unusual, the foreman insisted. Dasovich knew the dam was perfectly stable, it could never break, and it did no good to listen to these perpetual scare stories about a possible collapse. In a friendly way, Dasovich chided the deputy for butting in where he wasn't needed: Steve and his men could handle the situation without any help from the sheriff.

The two parted amiably, and driving their separate vehicles, started the long run up the hollow. They passed Sheriff Grimmett's home, at the top of Man, where the seventeen-mile-long hollow begins. They drove past the rows of darkened homes at Kistler; past the deserted parking lot of the big Island Creek Supermarket in Accoville; past the Buffalo Boy Scout Camp, with its huge, concrete activities building, at Crites. They drove on into Lundale, past the Lundale Baptist Church and the office headquarters of the Amherst Coal Company.

At Lorado, near the Gulf Oil gas station which faced the road in the middle of the town, Mutters noticed a number of people standing together in a group. They seemed to be watching the creek. The deputy pulled his patrol car over and got out. As he headed toward this group of early risers (miners mostly, waiting for rides to their shift in the nearby mines), Mutters saw Dasovich swing around his car and head on toward the dam, three miles above.

Mutters delivered the warning he had been sent to give, and quickly climbed back into the patrol car. His plan was to proceed to Three Forks, activate the blue flashing light on the roof of his car, and then make his way down the hollow, warning everyone he met and advising them to tell their neighbors of the danger.

When the deputy reached Three Forks, he found it mostly deserted. Most of the families were still in the

Lorado schoolhouse in the area which Mutters had just left. But Otto decided to make one stop anyway, at a small farmhouse located about half a mile below the little village. The farmhouse belonged to an elderly widow named Dellie Trent, who lived there with her three grown sons, her divorced daughter, and the daughter's teen-aged boy. Mutters figured the Trents probably hadn't heard the warnings yet. Their house sat right next to the creek in the middle of the narrow valley floor, and they would be terribly exposed if the water came.

Mutters parked the Ford and knocked on the door of the farmhouse. Mrs. Trent answered the door. "I told Dellie there was a good chance of the dam breaking," Mutters remembered later, "and I asked her to go. She wanted to leave. But she talked to her daughter (Wanda), and the daughter didn't think it was nothing to worry about. So Mrs. Trent thanked me anyway, and they stayed in the house."

It was a terrible mistake. Within two hours, Dellie Trent, her three sons, and her daughter would all be dead.

By 6:30 A.M., Mutters was back in Lorado. The first person he met there was fellow deputy Max Doty, who had just arrived on the scene to help with the warning process. The second person he met was Steve Dasovich, just returned from his 6 A.M. inspection of the threatened dam. Dasovich was still in his cheerful mood. Now he confronted the deputies as they sat parked together on the road near the Lorado schoolhouse.

"You can go on home now, boys," both later quoted Dasovich as saying. "We've dug a ditch. We've channeled around that thing, and the problem's all taken care of. There's nothing to worry about now."

The deputies looked at each other. If Dasovich's men had dug an overflow ditch—if the excess water was being allowed to drain quietly off—then it sounded like the situation was under control. After all, the big dam was Steve's responsibility he was supposed to know more about it than anyone else. . . .

Of course, neither deputy had traveled up to the huge impoundment to see for himself how dangerous it looked. But would they have even known what they were looking at? The whole thing was perplexing as hell, even a little ridiculous. It would be broad daylight soon. People were already walking around, going in and out of the nearby schoolhouse. The deputies would look pretty silly, playing Paul Revere up and down the hollow, after the company vice president had already eliminated the danger once and for all.

Max Doty made his decision quickly. Deciding that Dasovich probably knew what he was talking about, Doty left the scene and went to breakfast at a restaurant he owned in Kistler—Doty's "Shake and Burger." During the hour and a half left before the catastrophic break in the dam, Deputy Doty would warn no one.

Mutters was not quite convinced, however. After leaving Dasovich the deputy began chatting with several of the older miners who lived around Lorado. They told him that Dasovich didn't know what he was talking about, that the dam was still in danger and might break at any moment. Mutters decided to continue the warning. But he admits that, after talking to Dasovich, his efforts were half-hearted, carried out at random. (None of the dozens of residents interviewed after the disaster remembered being warned by a sheriff's deputy.) "If Steve hadn't told us that the dam had been fixed, that it was in good shape, I don't think more than a dozen people would have died," Mutters said later.

By 7 A.M., Dasovich had gone to the Buffalo Mining Company office in Lorado to begin his day's work; Max Doty was sitting at the "Shake and Burger" drinking cof-

fee and Otto Mutters was making his leisurely way down
the hollow, warning occasional passers-by to keep their
eyes and ears open. The critical warning—the one which
might have saved more than 100 lives—had failed. The
precious, five-hour reprieve provided by Mrs. Adkins had
been thrown away, wasted. In sixty minutes, the nightmare would begin.

The people in Three Forks were frightened to death,
the sheriff's department was all up in arms, and Steve
Dasovich couldn't figure it out.

Soon after receiving Kent's warning call on this wet
Saturday morning ("Steve, there is an awful lot of water
behind the impoundment," the strip superintendent had
told him during their 5 A.M. telephone exchange), Dasovich had traveled up to the big Three Forks dam to assess
the danger for himself.

And just as he had suspected, the situation looked perfectly safe. Sure, there was a lot of water behind the massive dam—in fact, the water level was now within a foot
of the crest. But that was nothing new. The water level
always rose during periods of heavy rain. Dasovich had
never seen it quite so high before, but he was positive that
the enormous barrier would hold. Anyway, even if the
water topped the dam a little, it would just run out in a
small stream, a brief overflow, before the rain finally
stopped and the water-level once again receded. The
dam was more than 400 feet wide in most places, and it
would take a hell of a lot more water than they were
looking at to move it out.

But if the dam were safe, why was everybody so ex-
cited? Dasovich supposed it was the same old story—
well-intentioned people, probably, spreading rumors and
exaggerated reports they'd heard at third-hand somewhere. These scary stories about an approaching break in

the dam were nothing new. Dasovich had heard them
himself on more than one occasion.

But Daniel S. Dasovich had not gotten to where he was
today by sitting around trading silly rumors he knew
nothing about. Not by a long shot. The vice president of
the Buffalo Mining Company—the top official on the
scene in West Virginia—was a self-made man who had
worked his way up through the ranks. While still in high
school, the stocky, sandy-haired Dasovich had gone down
into the mines as a common digger, and learned the trade
from the bottom up. After a brief stint in the service
during World War II, he had returned home to put himself through West Virginia University on the G.I. Bill—
earning a degree in mine engineering in 1949—and had
then gone back to work for the coal companies where he'd
made his start.

All in all, Dasovich had more than twenty-five years of
mining experience. Joining the Buffalo Mining Company
in 1967, he had quickly demonstrated the driving aggres-
siveness and the fierce desire for ever-increasing production figures which were his trademarks. Steve Dasovich
got the coal out—that was one thing they could say about
him for sure.

Of course, there were miners about Buffalo Creek
Hollow who said a lot of other things as well. Some of
them described Dasovich as a ruthless man who was only
too willing to take chances on safety in order to meet the
next production quota. Others called him arrogant, a
proud man who could not take criticism, who had to do
everything his own way.

"Steve has his own ideas on everything; he won't listen
to anybody else," was a common criticism of the Buffalo
vice president around the hollow. But everyone agreed
that Dasovich was smart, that he was tough, and that he
always produced the required amount of coal.

As he stood surveying the dam with Jack Kent, Dasovich was convinced that the situation was under control. Still, he might as well stay on the safe side. Talking the problem over with Kent, the boss decided that the best way to eliminate the irritating possibility of an overflow was to install a small drain pipe in the surface of the dam. The pipe would be placed across one corner of the dam at an angle, and would empty into a roadside ditch which ran parallel to the impoundment on the right side of the hollow looking downstream. "Jack," Dasovich told his strip superintendent, "I guess the thing we need to do is put us another piece of pipe in here." With these words, the boss gave his go-ahead for the project. Kent was to round up a crew of construction men, and a welder to fuse together two long sections of 24-inch, steel pipe. Meanwhile, the construction men would prepare the ditch for the pipe's placement.

Leaving Kent to work out the details, Dasovich headed back down the hollow toward his office. At the Lorado schoolhouse, he came upon the two sheriff's deputies who had supposedly been sent up Buffalo Creek to spread a warning about the dam. He had news for them: the problem was solved, the dam had been repaired, they could head on home now. The deputies seemed to accept his explanation and did not question him.

When he reached the small, white-painted Buffalo Mining office, Dasovich decided to make a phone call. There was no danger, he knew, but just for the record, he figured he'd better call the big boss over in Virginia, and let him know what was going on at the dam. Dasovich picked up the telephone, and quickly dialed the number for I. C. Spotte.

Spotte, the president of Buffalo Mining, was also the top-ranking officer for The Pittston Company's area coal group, which included several of the New York conglomerate's coal-mining subsidiaries in Virginia and West Virginia. Headquartered in Dante, Virginia, Spotte was the link between the little coal companies out in the field and their parent corporation in New York.

When Spotte came on the line, Dasovich told him about the problem at Three Forks. He explained that the situation was perfectly safe, but that a lot of the local residents were worried about the dam. Dasovich told Spotte that because he "wanted to relieve their [the residents'] anxiety," he was going to go ahead and put an overflow pipe in. Spotte approved of the decision—installing a drain pipe in the dam probably would make the frightened residents feel a lot better—and told Dasovich to keep him informed of the situation as it developed. The two then hung up.

By 7:45, Steve Dasovich was down in Lundale, at the Island Creek Supermarket located in the middle of town, hoping to buy several raincoats for the men who would soon be working on the dam. Along the way, he had stopped to assure half a dozen people that the dam posed no threat whatsoever. He advised the miners who were supposed to work that day to go ahead and put in their shifts, because the mines would be operating as usual.

When the roaring, deadly surge of water ripped through Buffalo Creek Hollow, Dasovich would be on the road at Craneco (just above Lundale), and he would have to scurry for his life. He would be "just amazed" to see houses slamming together, whole sections of railroad track whiplashing through the air, dozens of people dying on every side. He would be amazed that he could have been so wrong.

And by then, he would be powerless to help.

The windshield wipers shuttled back and forth, to a clacking, sing-song rhythm, and sitting at the wheel of his bright green Cadillac Coupe De Ville, Denny Gibson gave a deep, bone-weary yawn.

It had been a long night. Like the other residents of Three Forks, the thirty-nine-year-old Gibson had known well before midnight that the huge dam, located only a few hundred yards above his town, was perilously close to breaking. Like most of the others, he had hours ago decided to move his family down to the Lorado schoolhouse for safekeeping. Then, as the long, wet night slowly passed, he sat in his living room at Three Forks, working on a set of election posters which would be needed during his upcoming campaign for the office of constable.

Twice during the night, Gibson had left his posters and the warmth of his home to journey up to the dam with his boss, Jack Kent. Gibson was not comforted by what he saw at the dam: the water was rising with frightening speed.

Only a few minutes before, Gibson had driven the green Cadillac down Buffalo Creek Hollow to check once again on his family, to make sure they were getting along all right in the schoolhouse. They were. The grown-ups sat drinking coffee, or dozing; the children scampered up and down the brightly-lit hallway, in and out of empty classrooms. Now, as he drove back up the hollow road, around 7:30 A.M., the squat, moon-faced Gibson was heading for one last look at the threatened dam. After that, he hoped to catch up on some badly-needed sleep.

Peering through his windshield, Gibson piloted his Cadillac along Highway 16, past the tiny village of Three Forks, past the Free Will Baptist Church at the very top of the settlement. Just above the church, Gibson turned onto the narrow haul road which ran up to the dam and the mines behind it. By now the dawn had broken, pale as ashes, under a sky running with silver-black clouds.

When Gibson reached the impoundment, he parked his

car and looked around. The place was deserted—only the frigid, late-winter wind came whistling across the dam's vast absolutely flat surface. He opened the car door, took a few steps out onto the top of the dam, and suddenly sank up to his ankles.

"Lord have mercy," Gibson said to himself, "this thing's done gone soft!" He took another step, and again the wet muck covered his shoes. Something was definitely wrong. Only a few hours ago, he and Jack Kent had walked all over the dam, checking the measuring stick, examining the crest from one side of the valley to the other. Now, it was almost impossible to move around. Gibson stood absolutely still. He watched. Slowly, right in front of him, water was seeping up between the piles of gob which had been dumped on the dam's surface for eventual grading. Just as Jack Kent had predicted, the water was beginning to top the Three Forks dam. Or else the huge barrier had started to sink into the water behind it—Gibson couldn't tell which. Hurriedly, he looked for the top of Kent's measuring stick, which should have been visible from where he was standing.

It wasn't. The stick was gone. The water was lapping the top of the dam. And the little, protruding island of gob and slate, on which he and Kent had stood while checking the stick, was slowly dissolving. Gibson jumped back in the Cadillac, wheeled it around, and raced for his family in Lorado. He was the last person who would see the dam still standing.

They were joking, Johnny Wells remembers, he and Warren Adkins and some of the other miners, as they worked the midnight, or "hoot owl," shift at Buffalo Mine No. 5, located about half a mile above the Three Forks dam. Joking—and kidding each other about what would happen when the dam below them broke. "You'll never get home," Johnny told Warren with a laugh. "That flood will wash out the roads and you'll have to walk out of this valley on foot!"

It had been a strange shift, with less than half of the miners showing up for work because the heavy rains had blocked roads all around Buffalo Creek Hollow. But there was always something to be done inside a coal mine. Wells remembers that his undermanned crew spent most of that Saturday morning's shift moving a mechanized belt and a motor from deep inside the mine to a repair shed at the surface. As they worked, the miners talked about the reports of trouble at the nearby dam. The water was up higher than most of them had ever seen it, but nobody really expected the dam to break.

Wells wasn't overly concerned, anyway, because his own family lived several hundred yards up Davy Hollow (a small, side-valley which funnels into Buffalo Creek at Lorado), and they were safely out of the danger area. As the hoot owl shift ended, around 8 A.M., and Wells washed up with his fellow miners in the bathhouse, he kept up a running stream of jokes about the threatened dam and the flood that was to follow. Then he climbed into his car and headed down Middle Fork toward home.

Wells rounded a curve, with the dam perhaps 200 yards to his left, and suddenly he was watching an enormous explosion cover the impoundment "like an atomic bomb." Stunned, Wells hit the brakes. The car skidded to a stop, and within seconds, its front end was completely covered with a thick layer of black, gooey muck. "I couldn't tell what it was," Wells said. "I couldn't hardly see. It looked to me like the dam had blowed up, and had blowed that dust back up on the hill."

Actually, Wells was watching a series of explosions which occurred as the millions of gallons of water, after cutting through the collapsing dam, careened into the

200-foot-high, burning gob pile below the impoundment. Like water poured on the glowing coals of a campfire, the surging flow sent clouds of steam and ash roaring hundreds of feet into the air.

Wells couldn't believe his eyes. After all his joking, the Three Forks dam was being ripped to pieces in front of him. Frantically, he whipped his car around and sped back to the mining office he had just left. Somebody had to get on the telephone and warn the hollow of what was coming. But as Wells approached the mine, he saw its lights flicker off. The power was gone now, and there was no way to warn anybody.

Ossie Adkins, retired coal miner, gnarled and skinny as a stick, was just sitting down to breakfast with his family. It was a big breakfast—sausage, eggs, biscuits, several kinds of fruit—but Ossie didn't get to eat much of it. He had taken only a few bites when he heard two things at once: his daughter's voice, yelling for him to come outside quick, and a series of explosions "like blasting powder going off."

The Adkins family lived closer to the Three Forks dam than anyone else—in a small, white-painted home at the mouth of Lee Fork Hollow, which (along the Middle and Main Forks) meets Buffalo Creek near the burning gob pile. Ossie's front porch looked directly out on the back of the gob pile, perhaps fifty yards away. Now, as Ossie and his startled family stood on the porch, the huge pile seemed to have become a volcano. Great funnels of white smoke shot through the air; one after another, explosions of steam shook the ground. Moments later, a wall of water estimated at more than fifty feet high knifed through the last of the burning gob and hurtled toward Three Forks.

The water was loose now, approximately 135 million gallons of it, about to begin its seventeen-mile-long run to the Guyandotte River at Man. In its half-mile plunge from the top of the coal-waste dam to the floor of Buffalo Creek, the floodwater had already fallen 250 feet—a drop which easily surpasses Niagara Falls. As it traveled the valley for the next three hours, it would fall another 600 feet, developing enormous pressures and shattering most structures in its path.

Before the waters finally dissipated into the Guyandotte, they would kill 125 people, wreck more than 1,000 homes. Property damage would exceed $50 million—with an estimated $15 million in highway damage alone—and more than 4,000 people would be left homeless.

The clock on Ossie Adkins's wall had stopped at exactly 8:01 on the morning of February 26, 1972. That was the minute in which it began, the worst flood in West Virginia's history, the latest in a long, seemingly inevitable line of disasters which history shows are a part of living in West Virginia, a part of mining coal. In the desperate hours ahead, the great flood of 1972 would produce a hundred heroes, a hundred cowards, and change forever the lives of the thousands whose destiny was to have lived on Buffalo Creek.

Deaths

Three Forks

GAINING SPEED NOW, roaring along at close to twenty miles per hour, the advancing floodwave knifed through the last of the burning gob pile and plunged into Buffalo Creek Hollow. The first structure in its path—a Buffalo Mining Company garage—shattered to pieces as the wave struck, sending huge trucks and other pieces of heavy equipment flying through the air. Like a bucket of water poured into a child's sandbox, the torrent shot across the narrow valley floor, climbed almost forty feet up the opposite mountainside and began its deadly run down the hollow. A few seconds more, and the fifty-foot wave had reached the Free Will Baptist Church just above the village of Three Forks. It lifted the church right out of its foundations, flipped it end over end, and sent it skittering downstream.

Less than half a mile below the church, sitting in the living room of his company-built, green-shuttered home, forty-five-year old Leroy Lambert had just finished eating two fried eggs and a piece of toast. Now he lit up a cigarette, settled back in his easychair, and tried to think. Should he stay in his house? Or head for safety on the nearest hillside? Was the Middle Fork dam really about to break like everybody said? In five seconds, Lambert would have all these questions answered for him, as he ran for his life.

All night, Leroy's neighbors in the village of Three

Forks had been warning him to leave his home and seek safety in the Lorado schoolhouse, three miles down the hollow. The neighbors were frightened. One by one they had knocked on his door during the night and told him that the huge dam above them was about to break. If the dam ever went, the neighbors knew that their homes— sitting completely exposed on a valley floor less than 100 yards wide—would be destroyed. Pack up your belongings, the neighbors were saying, and get your family out while there's still time.

But Leroy Lambert, a stocky, one-legged coal miner with a voice like gravel on sandpaper, had never been one to rush off following somebody else's advice. If anything, the opposite was true: Leroy liked to lead the way. Whether it was drinking more moonshine whiskey than anybody else (two quarts at one sitting, he claims proudly), or being the first hunter along Buffalo Creek to try to shoot a bear with a bow and arrow, Leroy Lambert liked to be out in front.

It had always been that way. The son of a lifetime coal miner who raised nine kids on wages that sometimes ran as low as $39 a month, Leroy had as a child weathered the hard times, the relentless grind of poverty familiar to thousands who worked the West Virginia coal mines during the Depression era. Frank Lambert had not been able to provide many material goods for his children, and Leroy Lambert vowed that when *he* grew up, he would enjoy all the good things life had to offer.

And he kept his vow. During his twenty-nine years in the coal mines around Buffalo Creek, he had not been content simply to put in his shift. Between regular, eight-hour stints in the mine, Leroy had done everything from short-haul trucking to selling household products door-to-door. The extra work had paid off: Lambert was proud of the new aluminum siding—$4,800-worth over the last several years—which gave his once-drab house a prosperous, classy look. He was happy with the new bathroom he'd installed, the wood-paneled addition to the basement, the new wire fence and gate around his house, and the lush grape arbor out back. His big color television set, his new stereo with more than 400 albums, his cars and trucks—all of these were convincing proof that Leroy Lambert had made the grade in life.

Best of all, Leroy knew that his own family—his plump, dark-haired wife, Easter, and his two sons, Roger Dale and Anthony—would never go without material goods in the way that he had gone without them while growing up. These days, the Lamberts lived in a comfortable, well-furnished home. They took time for the fishing and hunting which Leroy and his sons loved. They enjoyed listening to Leroy's huge collection of hillbilly quartet records (Leroy had been something of a singer in his own right, back in the years when he took the "Lambert Quartet" down to Station WLOG in Logan for weekly radio shows).

All in all, with Roger Dale newly-married and living with his parents while he looked for a home of his own, and with Anthony safely returned from the Vietnam War, life seemed as promising and prosperous as it could get. Even the three dogs—the Pekingese pups Nan, Suzie, and Mia—seemed to be enjoying the good times. Plump balls of fur, they lay perpetually motionless on tables and chairs, awaiting the next call to dinner.

Leroy enjoyed his possessions. He also felt he'd earned them. Lambert knew better than most how dangerous it was to make a living in the coal mines of West Virginia. Half a dozen times, he figured, he'd come close to dying in the underground gloom where he earned his paycheck. The worst accident, of course, had been the one at Amherst Coal's No. 2 mine in 1965. Sometimes, remembering that horrible afternoon at No. 2, Leroy imagined he could still feel the pain in his leg; could still see the big coal buggy rumbling toward him, crushing him against the seat of the Super-14 Joy coal loader on which he sat helpless. The buggy driver had lost control for a moment, Leroy remembered, only an instant, but enough time to send his machine crashing against Lambert's loader, pinning his leg and mashing it to pulp. After that, there had been an agonizing, two-hour wait for the ambulance at the mouth of the mine, and then a forty-five-minute ride to the hospital. When he finally arrived, the doctors told Lambert that the tissues in his crushed left leg had died, and took it off at the knee.

After his accident, Amherst Coal had given Leroy a new job, a job even a one-legged man who hopped around on an artificial limb could handle. Now he worked as a bathhouse attendant outside one of Amherst's mines. Each day, he hosed down the bathhouse where the miners washed after their shifts, then cleaned and repaired the dozens of safety lamps which stood in rows out back. The job paid $36.05 a shift, there was no foreman to stand over your shoulder all day, and Leroy liked it.

Lambert's case was not unusual. All around him, there were similar examples of the toll a life in the mines can take. His father, Frank, nearly blind from his years in the mines, had finally died of silicosis in 1956. His wife's father, Aubra Peters, had also died of silicosis, in 1952. His brother-in-law, Enoch Peters, had perished in a slate fall four years earlier. His daughter-in-law's father, Bill Imes, had also been killed in a slate fall, in 1950. Two of Leroy's uncles had died of coal mine-induced silicosis, and a cousin had been crushed to death in yet another slate fall. The evidence was staggering: coal mining was surely the most dangerous work on earth.

Long accustomed to living with danger, then, Lambert was not about to panic when his neighbors told him that the Three Forks dam was in trouble. After all, these warnings were hardly unusual; they came like clockwork each spring, when the hard mountain rain fell in torrents and water rose in every creek. Leroy listened to the advice—and went his own way. He would remain awake during the night, listening to flood reports on the radio and trying to assess the danger, but he would not leave his home until it was absolutely necessary.

And so they sat, Leroy and Easter, twenty-four-year-old Roger Dale, who worked as a coal washer at the Buffalo Mining tipple near the dam, and young Anthony, just back from Vietnam and still unemployed. They finished their breakfasts and one by one wandered out to the living room, where the three dogs slumbered and the radio played on, periodically interrupting the morning's music with up-to-the-minute flood reports. Around 7:45, they heard a knock at the door. It was Easter's sister, Ethel Sparks, Ethel's married daughter, Wanda Carter, and Wanda's one-year-old baby, Matthew. Ethel made her point quickly: she wanted Leroy and his family to leave Three Forks, the danger area, and follow her to her own home at Elklick, about a mile and a half down the hollow. The guests sat down, and the Lamberts began to discuss this plan.

After a few moments, Easter remembered that she had been in the middle of making herself a bowl of Corn Flakes when the visitors arrived. Now she walked back into the kitchen. She poured the cereal into a bowl, then the milk, then the sugar. She looked out the window and noticed two things at once. First, the two-day storm seemed to have gotten worse. As she peered up the hollow road that ran toward the dam, the sky looked pitch-black, darker than she had ever seen it.

Second, Freddie Thomason, a neighbor who lived in a trailer only a few yards away, was acting awfully funny.

He was standing out in his yard, jumping up and down, waving his arms back and forth. Perhaps he was yelling hello to somebody on the road. . . .

Walking back into the living room, Easter set the bowl of Corn Flakes down on a table, turned, and said: "Lord, look how dark it is up the road!"

At that instant, the lights flickered off and the radio went dead.

Everybody stood up. Leroy looked at the clock: it read 8:01. "The power's kicked off," Leroy said to no one in particular, and then he noticed something that made him jump. Through a living room window, he could see the power lines attached to his house beginning to shake. Soon they were swaying back and forth, maybe three feet at a time, as though an enormous hand was tugging at them somewhere up the hollow. Puzzled, the family hurried out to the front porch. Freddie Thomason was still jumping and waving, only this time, they could hear the words:

"Run for your lives, here comes the church house!"

For a moment, nobody moved. They stared, not yet understanding, at the long line of houses above them, and then they saw it: an immense tongue of water, perhaps 50 feet high, a soot-black, foam-frothed wave of water barreling down the hollow straight at them. In front of the wave, the Free Will Baptist Church went bouncing along, tumbling over and over like a woodchip in a windstorm.

Three Forks was coming apart. House by house, every standing structure was being ripped out of the ground and hurled straight at them. Suddenly, the Lamberts were in a race with death.

Grabbing his crutches, Leroy yelled at the others to follow him, and led the way out to the family's 1971 red

Chevy Nova. But Easter and Tony had disappeared, had raced back into the house to save the three dogs. "They was in there chasing them pups," Leroy said later, "and we was screaming for 'em to come on—here come the church house through the camp!"

The wave advanced. Leroy screamed. By now, Easter and Tony had caught two of the pups. Desperately, Tony lunged for the third, but Suzie scampered away to hide under a table. They left her to perish, and stumbled back outside. As they dove into the Nova, Leroy hit the gas and gunned the car down the hollow. By this time, the wave was only 100 yards behind them, and water was running under the car.

Ethel, Wanda, and baby Matthew, meanwhile, had already started down the hollow in their own car. Up ahead, they had reached the highway bridge which crossed the creek about 100 yards below Three Forks. Suddenly, their car slammed to a stop.

"They stopped their car and both doors flies open at one time," Leroy remembers. "I didn't know what to think. I said, 'What in the world are they doing?' About that time, I look, the water was going forty feet over the top of the bridge. It was going down the creek thataway, and beat us there."

Racing down the creek bed, the floodwater had out-flanked Ethel and Wanda on the right. Now there was no way to cross the bridge. Grabbing the baby, they ran to the left, across the railroad tracks and up onto a small pile of gob which lay against the hillside. The flood was moving faster, swirling in a great circle as it rounded the curve near the bridge. Another few moments, and the Lamberts would be trapped in thirty feet of boiling water. "When I saw them run," Leroy says, "I knew I wasn't going to have time to stop. Because by that time, the church house was almost pushing up against me. So I whipped the car straight across the railroad tracks and shot the gas to it. That was the only chance we had."

The car jumped forward—but then, for a heart-stopping second—seemed to have stalled on the tracks. The back wheels spun helplessly, then caught, and the car slammed head-on into the gob pile which Wanda and Ethel had just climbed. Now the Lamberts were at the bottom of the thirty-foot pile with the water closing in on them. They jumped out of the Nova and tried to scramble up the steep sides of the gob pile. ("So we started up the slate dump and my wife took about two steps and she slid down and went into shock. She already had high blood.")

Completely panicked, Easter lay with a pup under each arm, her face buried in the muck, her elbows digging into the wet gob. Leroy had lost his crutches. Now he crawled along the slick, rain-soaked dump, looking back at his wife in anguish. Finally, Anthony jerked the pups from his mother's arms, and hurled them thirty feet through the air, high up onto the gob pile. Sliding, stumbling in water up to their waists, the two sons dragged their mother bodily up the slope.

One danger followed another now, almost as fast as they could react. Suddenly a huge, water-borne trailer was bearing down on them; just as suddenly, it sank out of sight. Leroy screamed to his sons: "Don't pull her, jerk her! Get her out of there!" For a few seconds, they struggled with their lives in the balance. Rain-soaked, covered with gooey sludge from the failed impoundment, they gasped and choked their way up the side of the refuse pile. "The water took her shoes off, it like to tore one leg off her. And I was scared the water was going to cut the slate dump out from under us. But she held, and we made it."

Shivering in the thirty-degree temperature, they stood and watched the water catch the Chevy they'd just left; watched the wave vault the car fifty feet in the air as it came barreling through. There were horrors yet to come: a moment later, they heard a voice screaming for help, and then stared helplessly as one of Jason Bailey's young children (the only one who would survive, as it turned out) went flying past. Held uncannily aloft by the water's furious whirlpool, the boy raced downstream. He yelled for help again and again as he passed. In a moment, he was gone. (Later, miraculously, he managed to grab a tree branch and pull himself out of the water, almost unscathed.)

A minute later, the rain had turned to snow. The Lamberts stood on top of the gob pile while the flakes fluttered past, strangely peaceful, riding silently on the frigid wind. Everything seemed unreal. Thirty feet below, Three Forks had broken into a thousand pieces and washed away. They stared numbly down on the place where they'd spent most of their lives. It was gone without a trace.

By now, Easter had fallen into acute hysteria. All she could see was the black, greasy water; all she could hear was the roar of the houses breaking up. The men began collecting pieces of wood, in the hope of building a fire. But each time one of them moved more than a few feet away from Easter, she began to scream again, terrible, wrenching screams that echoed in the hills above Buffalo Creek. "I thought she was going to die right there on the slate dump," Leroy says. "She was all tore up, her back was hurt, and she kept screaming and screaming and there was nothing we could do about it."

Finally, the agony ended. Ossie Adkins, climbing along the hills above the hollow, found the Lamberts standing together on a hillside, shivering in the wind. He led them back to his own, un-harmed house, to the warmth of blankets and hot coffee and food. Easter was seriously injured (she would spend seven days in the hospital, her back in traction), and the Lamberts had lost all of the material possessions of which Leroy had been so proud. But they were lucky. They were still alive.

Others were not so lucky. At least twenty people died in and around Three Forks—most of them the members of five different families who had remained in the area, and who were trapped when the sudden flood came surging through. All the homes in the immediate vicinity were gone. The Free Will Baptist Church was gone. Survivors noted later that the only landmark left in Three Forks was one lone weeping willow tree, still standing in the middle of where the village had been.

By now, as it plunged out of Three Forks and began its three-mile run to the town of Lorado, the flood had picked up a set of deadly teeth. Shattered timbers, sections of railroad track, millions of tons of coal-waste debris—they sliced back and forth inside the descending wave. For many who waited in the towns below, this tumbling wreckage would prove more dangerous than the water itself.

Less than a mile below devastated Three Forks, seventeen-year-old Hebert Trent had just strolled out to his front porch to take another look around. Stretching, he sniffed the damp morning air and the smell of his grandmother's bacon, frying on the stove back in the kitchen. It was going to be a big breakfast because the Trents had been up most of Friday night worrying about Buffalo Creek and wondering if the dam at Middle Fork was going to break.

There had certainly been enough warnings. Several times during the night neighbors had stopped to knock on the door of the Trent farmhouse, located only a few yards from the creek. The worried visitors had advised

Dellie Trent to get to higher ground, but the elderly coal miner's widow had talked the situation over with her divorced daughter, Wanda Osborne, and in the end, had decided against leaving. Still, it was hard to sleep knowing about all that water behind the impoundment at Three Forks. On several occasions during the long hours of darkness, Dellie's three grown sons—all of whom were coal miners and all of whom still lived at home—took turns going outside to check the creek.

The Trents lived in a farmhouse, with three small barns above them and a detached garage about thirty yards below. The family kept seven cows and a horse in the barns; the garage was occupied by a ton-and-a-half, red Chevrolet truck in which Henry, one of Dellie's three sons, hauled hay and feed for the animals. It was a small, but sufficient farming operation. The rows of corn, cabbage, tomatoes, potatoes, and onions below the house provided the Trents with all the vegetables they could eat.

Hebert had not slept very well during the night. Every few minutes, it seemed, somebody else was pounding at the door to warn Dellie of the growing danger. After each of these alerts, Hebert's uncles—Gene, Johnny, and Henry—would steal outside to take yet another look at the nearby creek. It was rising fast, but didn't seem to be unusually high for this time of year.

The most ominous note had been struck around 6 A.M., when a sheriff's deputy suddenly appeared at the door. Deputy Mutters told Dellie that the sheriff had received several calls about the Middle Fork dam; he suggested that the old lady, her three sons, her daughter, and her grandson move downstream to the Lorado schoolhouse, where a lot of people were gathering for safety. Dellie said no; it would be light soon, her three sons were watching the creek, they would have time to run if the water came.

After the long night of vigilance, the problem at the dam had finally become something of a joke to the Trents. Hebert remembers how they were laughing about the situation: "That thing's never going to break," chuckled one of his uncles, "but that's all you hear around here—look out, the dam's ready to go!"

Now, just before settling down to his big breakfast, a hungry Hebert Trent had wandered out to the front porch to take one more look. As he stood on the porch, staring up the hollow, he was surprised to see the power lines begin to shake. They were twitching, jumping back and forth to a regular rhythm, and Hebert had never seen power lines act like that before.

At the same moment, a blue Ford pickup truck came roaring down the hollow road, maybe twenty-five yards away. The truck's flashers were going, and the horn blared again and again, shattering the silence. As he shot past the farmhouse, the driver leaned far out his window, and Hebert heard the words: "Get out of there, the dam has broke!"

Instantly, the youth turned and ran through the living room, shouting, "Run for it! The dam's broke! The dam's broke!" The others looked up at him, stunned for a moment. Hebert kept running. He didn't know where he was going, but his first thought was for Joe, his favorite coon dog, still chained to his box on the back porch. Slamming through the back door, Hebert unsnapped Joe's chain and dragged him by the collar across the yard. His plan was to put the dog inside a pickup truck parked out back, then climb in himself and ride the flood out.

Hebert had opened the back door of the pickup, and was lifting Joe inside, when he looked up for a moment. Glancing through the truck's windshield, he saw the oncoming wave bearing down on him. The water had reached a barn above the house, and it was already level with the roof of the barn, about twelve feet high. Hebert jumped. "I just dropped everything and took off through the yard," he says. Joe landed back on the ground, where he would drown in a few moments.

Hebert hardly knew what he was doing, but his instinct for survival was sure. Racing into the garage below the house, he dove onto the back of Henry's red Chevy truck. With all his strength, he clutched at the wooden carrying racks in the truck bed. He lay face down, his eyes shut.

The water slammed into the garage, blasting it to pieces. Suddenly the truck was airborne, turning and tumbling through space. His eyes still tightly closed, Hebert hung on desperately while the truck toppled end over end down the crashing wave. So far, he could still breathe—the truck was rolling so fast he never remained underwater long. The churning flood tore Hebert's trousers to shreds; it ripped the bottom off one of his shoes. Around and around he went, in the soot-black water. Every few moments, the truck slammed into another obstruction. "I just knowed I was going to die," Hebert says, "but I was trying to hold on as long as I could."

About a third of a mile below the house, the water was backing up behind a railroad trestle. As the current slowed for a moment, Hebert suddenly found himself back at the surface. The roof of a house floated nearby; somehow, he fought his way over to it. The ride was gentler now, gliding down the valley on the roof, but Hebert had no way of knowing how long it would last. Rounding a bend, he saw a tall stand of trees hanging out over the water.

In a moment, Hebert decided to take the chance. He would jump. If he missed and fell back in the water, he would almost certainly drown. But it was a way out, if he could make it. Wiping the mud from his eyes, he gathered his strength and jumped straight up—perhaps five feet—

from the floating roof to the heavy branches above him. Gasping, Hebert held on for a few minutes to catch his breath, then pulled himself over to the trunk.

He was cut and scratched, and every muscle ached. But Hebert had lived ("I just knowed I had somebody with me," he says, "somebody I never had before"). He limped down to Lorado, where friends gave him a pair of dry pants and a new pair of shoes. A few hours later, after he had recovered from the beating of his mile-long ride, Hebert hiked back to Three Forks. He hoped desperately that his three uncles, his mother, and his grandmother had found a way to escape. If any of them had gotten out, he figured he would find footprints in the soft earth around the house.

There were no footprints. All of them were dead.

Twenty miles to the north, at the popular Smoke House Restaurant in downtown Logan, radio newsman Bill Becker was just sitting down to a well-deserved breakfast. Becker was exhausted. As often happened during periods of flood danger, he'd spent the entire night on the air broadcasting weather bulletins and flood reports as fast as they were received at his station, WVOW. Now that the crisis seemed to have passed, Becker was looking forward to going home and getting some sleep.

But he was not destined to sleep on this day. As Becker lifted the first mouthful of food to his lips, he was startled to see one of his assistants, young Marty Backus, dash through the door of the Smoke House and hurry over to his table.

In a moment, Backus explained the situation. A man named Garland Scaggs—an office employee at the Amherst Coal Company headquarters in Lundale—had just called to say that the Buffalo Creek dam had broken. All hell was breaking loose, according to Scaggs, who described "A wall of debris riding down the valley, and people caught in the crest of the thing." Backus told Becker that it sounded like "One hell of a story," but that he didn't know exactly what to do with it.

Becker knew what to do. Leaving his unfinished meal on the table, he hurried up the street to his office at WVOW. Garland Scaggs was a good friend of his—the two had often played golf together—and Becker knew he would be able to recognize Scaggs's voice on the telephone. Of course, if the flood report wasn't a hoax, the phone lines might already be down.

Becker's call to Scaggs, which took place around 8:30, was one of the last telephone contacts made with Buffalo Creek Hollow that morning. And it was the first alert to the outside world that a tragedy of major proportions was developing in the hollow. Scaggs described a woman and a young boy, whom he had seen swallowed up in the flood. He described rows of houses stacked crazily on top of each other. The water had already reached Lundale, Scaggs said, but if Becker moved quickly, he might still save many lives farther downstream.

Becker moved. His voice, strained with tension, was the one which first brought the news of the disaster to thousands of residents across Logan County. "This is Bill Becker," the voice said again and again. "The dam at Lorado has broken. Get out of low-lying areas immediately. This is Bill Becker . . ."

CHAPTER 4

Lorado

Only a few minutes before Becker's dramatic broadcast, another attempt to warn Buffalo Creek Hollow of the impending danger had fallen tragically short. After receiving several telephone calls from frightened hollow residents who insisted that the Three Forks dam was about to break, Sgt. Doug Williamson of the National Guard's Logan unit had decided to check the reports out. (The Guard was operating a skeleton crew at its armory on this Saturday morning in response to the widespread flooding across southern West Virginia.) He quickly telephoned a close friend and fellow Guardsman—a coal miner named Billy Aldridge who lived at Lorado—and asked him to take a ride up to the threatened barrier. If the situation looked as bad as everybody said it was, Aldridge was to begin warning the public immediately.

Sitting on his end of the phone in Lorado, Aldridge stretched and groaned. He was tired. After putting in his Friday night shift at the Buffalo Mining Company—where he worked as a "slate picker"—the twenty-three-year-old Aldridge had sat up half the night listening to flood reports on the radio. A sergeant in the National Guard, he half expected to be called out for emergency duty as the floodwaters rose around Logan County. But the flooding had not been very severe, and during the early hours, Aldridge had finally drifted off to sleep.

By now, Aldridge's attractive young wife Jeaneil was also up, padding around the kitchen as she prepared breakfast.

Aldridge climbed into his blue jeans and grabbed a red and black-checkered hunting jacket from the closet. While they ate a hurried breakfast, Billy and Jeaneil discussed the danger. They couldn't conceive of the dam's actually breaking, but they wondered how much water had backed up behind it.

Ten minutes later Aldridge was rattling up the hollow road in his blue Ford pickup. He noticed a lot of people were up and about in Lorado; that seemed unusual for so early on a Saturday morning. Above Lorado, though, the winding road which ran alongside the creek was deserted. Aldridge peered through the deep gloom. It seemed awfully dark for 8 o'clock in the morning. A few more turns and the road straightened out on the approach to Three Forks. Perhaps 100 yards ahead, he could barely make out the highway bridge which stood at the bottom of the village.

Something flashed. Aldridge craned his neck. Was he actually seeing this? Right in front of him, a red-and-white automobile suddenly catapulted seventy feet through the air. Behind the car, an immense wall of black water came crashing through Three Forks, exploding houses and spinning cars around like tops. Aldridge slammed on the brakes as he grasped the situation: the big dam had broken, and it was going to be worse than anybody had dreamed. ("It looked like something you see on the horror movie," he said later. "It wasn't real. I mean, I didn't believe it.")

There was an immediate problem, however. Trapped between the steep rock cliffs on his left and the creek on his right, Aldridge had no way to turn the truck around. So he did what he could: ramming the engine into reverse, he gunned the blue pickup backwards down the hollow.

Gears screaming, the truck pounded down the road,

more than 100 yards to the first driveway in which Aldridge could turn around. "All the time, I could see this
water coming." he says, "it was like an ocean. I was going
so fast I thought the transmission was going to blow up
under my feet."

At last Aldridge lurched into a driveway near the home
of Henry Peters, one of three houses which stood in a
cluster just below Three Forks. All three homes were in
the direct path of the water, and now the margin of life
or death depended on sheer luck: two of the families heard
Aldridge's racket—his horn-honking and his shouts—
raced outside, and escaped. The third family (Willy
Dempsey's) slept on. Within sixty seconds, three of the
Dempseys were dead.

Racing downhill, Aldridge turned on his truck flashers
and kept the horn going. His warnings would be extremely
brief, as he fled the front of the wave, and many of those
who failed to understand them would die. The Trent
farmhouse loomed up ahead. Before Aldridge reached it,
he was surprised to see two vehicles coming up the road
toward him. They were about to drive head-on into the
flood, which now roared out of sight behind the next bend.
Aldridge hit his brakes too late, honking furiously, and
the first car, a light blue Volkswagen, flashed seventy-five
feet past him before it stopped.

Inside the Volkswagen sat Albert Hedinger, the quality
control specialist at Buffalo Mining's tipple, who was also
Aldridge's former commander at the Logan National
Guard unit. Hedinger, presumably driving toward the
tipple to begin his day's work, was puzzled by Aldridge's
flasher and his blaring horn. Slowly, he climbed out of
the VW and turned to look back at Aldridge. There was
time for only one short scream. The last time Aldridge
saw Hedinger, the latter was standing with one foot in

the car and one on the road. Seconds later, he disappeared into the wave and perished.

The second driver was Jack Kent, in his pickup truck, heading back to the dam to supervise the ditch-digging project. Kent's reflexes were quicker. Hearing Aldridge's shouted warning, he spun the truck around and followed him down the hollow to safety.

Racing past the Trent farmhouse, Aldridge yelled a warning and roared on through an area called Sandy Bottom. Up ahead, he saw Eb Sparks standing in the yard outside his farmhouse. Frantically, Aldridge honked and waved. Thinking that Billy was in an awfully good mood this morning, Eb waved happily back. (But he saw the water in time, and escaped with his family.)

The next little community was Pardee, where the water would annihilate sixteen houses, wash away a concrete church building, and rip an entire electrical station out of the ground, thus cutting off electricity to hundreds of homes farther down Buffalo Creek. At Pardee, Aldridge passed a family standing on a porch. As he shouted, they piled into a waiting automobile and raced to safety. A few feet farther, he yelled a warning to another family and saw them scatter.

Now he was in Lorado, his own community. Aldridge kept the honking up. He yelled at a crowd of people standing along the railroad tracks, and they scurried for the hillside. When he reached his own home, Jeaneil was waiting on the porch and the water was perhaps one minute behind them. Aldridge found time to knock on two or three doors before dashing back to the truck and again gunning it down the hollow. He shot past the school (its occupants were already flying for the hillside) and the Lorado Supermarket. At the lower end of Lorado, lived one of Aldridge's closest friends, Andy Doczi, Jr., a superintendent at Buffalo Mining. The Doczis had several small children; Jeaneil guessed that all of them were asleep. Andy's wife was pregnant, they would be completely helpless, and suddenly Aldridge couldn't bear to leave them. He pulled the truck over to the side of the road. "If you see that water coming, you take off!" he told Jeaneil, and sprinted for the Doczi home.

The door was unlocked. Without knocking, Aldridge plunged inside. Within seconds, he had them on their feet. But the parents wanted to dress the children before taking them out into the cold. "You don't have time to put on no clothes!" Aldridge howled. "Let's get out of here!" Carrying the Doczis' baby boy, Todd, he led the way back outside. It was too late to go back to the truck: above them, the houses were breaking up, shattering apart one by one. "Everybody was shouting and screaming," Aldridge says, "and you could hear this awful noise—the houses crashing and everything."

The water was smashing through Lorado, but Aldridge wasn't done yet. Leaving the Doczis safe on a hillside, he ran back to the double row of houses in the middle of the hollow, where Edith Bucy was desperately pushing her husband Charles along in a wheelchair. Aldridge squatted down in the mud, and the crippled man climbed on his back. But Charlie was heavier than Aldridge had thought. After thirty feet or so he stumbled and fell into the muck. The water was getting deeper; the houses going down like they'd been bombed. It looked hopeless for a moment—and then the crippled man's son suddenly appeared to help. Gasping, in knee-deep water, they dragged Bucy to safety.

Aldridge's long, Paul Revere-like ride had been stretched to the limit: he ceased his final rescue effort at the last possible moment, and at great risk to his own life. Later, dozens of Buffalo Creek residents would agree that Billy

Aldridge was the hero of the 1972 flood, and the National Guard would give him a citation for bravery.

Now, Aldridge stood on the bank, watching almost all of Lorado break up and disappear. "It wasn't real," he says. "The water had cleaned everything out. It was high and black, and it was taking the houses and just crunching 'em. I couldn't believe it. I couldn't think. I said, this ain't happening to me. I could see it happening, but I was just numb, like I was drunk or something. I was just staring out into space, you know."

Now one horror quickly followed another. Aldridge saw a woman and two children sitting beside a bundle wrapped in a pink blanket. "Would you help me?" the woman asked, very politely. "There's a man here dying." (The man was Donald McCoy. He lost his wife and daughter, but lived after Aldridge rushed him to the hospital.) A few feet farther on, some of the residents had found a baby floating face-down in the water. Now they were trying their best to revive him, pulling the black muck from his throat with their fingers. (The baby belonged to coal miner Robert Albright, and also survived.)

An hour later, Aldridge found Jeaneil farther downstream, where she had driven the truck when he jumped out to warn the Doczis. For the next several days, Aldridge would spend most of his time working to help the residents of devastated Lorado. For now, he was happy enough to know that he and his wife had survived the disaster. And happy to know that, when the warning call came, he had answered it.

From the red-brick schoolhouse at one end of town to tiny Italy Bottom at the other, Lorado was breaking into a million pieces. By now, at least thirty of the town's estimated 500 residents were either dead or dying. All but about ten of its 125-odd homes had been ripped from their foundations and washed downstream. The Gulf station, the post office, two churches, and all of the schoolhouse but the gym were also destroyed. (The refugees at the Lorado school—including Mrs. Adkins, who had phoned in the first warning to the sheriff—escaped, fleeing the wave at the last moment.) The Buffalo Mining Company office was full of mud and water, but remained one of the few structures still standing. As the floodcrest hammered through Lorado, traveling at perhaps ten miles per hour, the stories of incredible escape and grisly death multiplied.

For Oldie Blankenship, a jut-jawed, fifty-five-year-old coal miner, February 26 was to have been just another day off. Another long, restful day in which to nurse the various ailments he'd picked up during a thirty-year career in the mines around Buffalo Creek—the bad back, the defective kidney, the inflamed prostate. A recent addition to the miner's disabled list, the stocky, slow-talking Blankenship was enjoying his days of rest: sitting around his home in Lorado drinking coffee, smoking an occasional cigarette, watching television.

He'd paid for them. The son of a poor Boone County farmer, Oldie and his eleven brothers and sisters had struggled through their early years, scraping along from one day to the next. The Blankenships had been able to grow most of the food they needed—while going without most other material goods. In spite of his missing left hand (lost during a shotgun accident at the age of fourteen), Oldie had been happy to land a job in the mines, beginning his long coal-mining career with the Logan County Coal Corp. in 1942.

He'd made a good living, working a variety of underground jobs, and after thirty years of labor, had saved enough money to realize a lifetime dream: in only a few weeks he hoped to be moving down to Tennessee, where he had recently purchased some property. With his Social

Security and his "John L." (miners' pension, named after
John L. Lewis, the UMW president who started it), Oldie
knew he would have enough money to satisfy his life's pas-
sion, the bass and trout fishing he loved so much. In
spite of his nagging, painful ailments, both he and his wife
Edith were looking forward to a peaceful, relaxed retirement.

Blankenship was only vaguely aware of the big dam at
Three Forks, and had no idea, as he climbed out of bed
on this Saturday morning, that the barrier was close to
breaking. Edith was still in bed. Oldie fixed himself a
cup of coffee and sat smoking a cigarette. After awhile,
he heard several cars racing up and down the road blowing their horns. This was hardly unusual ("They do that
every day in the week"), and Oldie paid no attention.

A few minutes later, Edith climbed out of bed and
began getting dressed. Oldie strolled into the bedroom,
put on a pair of pants and a shirt, then sat down on the
bed to tie his shoes. As he reached for the second shoe,
he looked through the bedroom window and was surprised to see a lot of water running down the road. Looked
like the creek had overflowed for sure. Frowning, he
reached up and pulled the chain on a small lamp above
his head. The light switched on. Then it flickered off.
Startled, Oldie pulled the chain again. Nothing happened.
The power was off. "I knowed something was wrong
then," he remembers, "and right after that, I looked out
the window and saw the big water coming."

More than twenty feet high, the floodwave was by now
only a few yards from Oldie's house. Suddenly panicked,
he ran to the screen door at the back of his home. Edith
was standing out in the backyard, accompanied, as usual,
by her little dog Bounce. As Oldie pushed the screen
door open and shouted for Edith to look out, he could
hear Bounce furiously barking.

Their eyes met. Edith shouted: "Oldie, Lord have mercy on us!" and the wave engulfed them. "I looked up and voom! The next house hit mine and knocked me back into the bathroom. The water looked to be about twenty-five feet high—we never had a chance. Just as she said that, why, everything went out and we went out with it!"

Edith and Bounce were swept to their deaths. Lying on the floor of his bathroom, Oldie realized that his house had broken loose and was floating downstream. A moment later, the walls of the bathroom were coming slowly together—just like in the nightmare most people have had at one time or another—and Oldie was being crushed between them. "The walls come together on my head 'till I knowed my head was going to bust," he says. "I said it was the end, I knew I was going to die. It had done come together so tight that I couldn't stand it and I couldn't move no which-a-way."

Then the house smashed to pieces and the bathroom walls released their hold. Helpless, Oldie was propelled straight up through the ceiling, through a break in the roof. As his head plunged through the thick insulation material, he remembers holding his breath so that the dust wouldn't make him choke and sneeze. Now he was loose in the water, in a sea of timbers, telephone poles, and tumbling automobiles. He flew down the creek bed, catching sight of a crowd of people standing on a hillside near the school as he passed. A moment later he was underwater, holding his breath for what seemed to be more than a minute before surfacing again.

"I knowed if I got my breath I was dead, and if I held it, I'd die anyhow," Oldie remembers. "So I didn't have no choice!" Boards and other obstructions slammed into him, striking terrible, bruising blows. His eyes burned from the gasoline released by the Gulf station which had collapsed near his home. Frightened at first, Blankenship now felt strangely peaceful. An intensely religious man, he had already accepted the fact that he was about to die: "People say when they get in a place like that, they think about praying. But I'd prayed before. I'd made things right with God before that. In the Bible it teaches you, 'Be ye always ready, because ye know not when Christ comes.' But if you're prayed up, and your sins are forgiven, then you know it. I never thought about praying, I knowed I was gone. But I didn't want to die with an arm tore off, or a leg tore off, or all cut up like that. Seemed like I wanted to die more quick and get it over with."

At the lower part of Lorado, in the area called Italy Bottom, Oldie caught sight of a floating trailer and quickly secured a hold on its roof. After a few minutes, the trailer swung around at the base of a hill and lodged for a moment. Oldie leaped to grab a tree branch which hung out over the water. He hung on, gasping for breath. When the current slowed a little, he managed to crawl over to a large rock and pull himself out of the water.

"I don't know if there's ever been anybody that nigh death and then get out of it," Oldie says. But he wasn't out of it yet—in fact, his ordeal had only begun. With his legs and back terribly bruised, he could not walk. So he began crawling along the mountainside. "I knowed that if I stayed there for a very long time that I'd freeze to death. It was real cold, spitting snow. The wind was a-blowing and my clothes was tore offa me."

For three and a half hours, Blankenship crawled around a mountainside, pulling himself along with bushes and tree branches, shivering in the frigid wind. Finally, he reached a house set high in the hills above the hollow, and the blessed warmth of a fire. He had no hopes for Edith. "I knew she was dead when I got out," he said.

He was right. The following Friday, he identified her body at the morgue.

While Oldie Blankenship fought for his life at Italy Bottom, forty-two-year-old Robert Lee Albright was struggling frantically to reach his family in Lorado. Albright, a soft-spoken, bespectacled coal miner, had finished his hoot owl shift at Buffalo Mining's No. 8B mine around 7:45 and then, as usual, had climbed aboard a mechanized coal belt for the mile-long ride out of the mine. The trip on the rumbling, electric belt would take about fifteen minutes, and then Albright would jump in his car and head home for breakfast.

After twenty-five years in the coal mines around Buffalo Creek, Albright knew every phase of the mining operation. Beginning as a novice coal loader in 1946, he had worked his way up through the ranks. By 1972, he was making $42.80 a day as a skilled electrician—and the bosses, respecting his obvious expertise, were allowing him to work almost as many hours of overtime as he wanted.

It was a good life, nothing like the terrible years of poverty which Albright had endured as a child. The son of a lifetime coal miner who raised eleven children on his skimpy wages, Albright knew what it was like to struggle along from day to day, always short of money. The Albright kids had not gone hungry ("Potatoes and beans, that was it!"), but Robert had quit school after the sixth grade mainly because "You didn't have no clothes decent to wear." As he learned the hard lessons of poverty, the young Albright swore that he would do a better job of providing for his own family.

He did. Nowadays, the Albrights—Robert, his wife Sylvia, his seventeen-year-old son Steve, and his newly-adopted baby, Kerry Lee—lived in a warm, well-furnished home in upper Lorado, a few dozen yards above the schoolhouse. The family drove two cars, enjoyed a houseful of new appliances, ate steak whenever they wanted it. And in a few months, Robert would be sending his son, a talented saxaphone player, off to study music at Fairmont State College.

All in all, Robert Albright considered himself better off than he had ever been. He had suffered some terrible disappointments along the way, of course: the death of his oldest son (killed in Vietnam in 1970) was still a daily source of pain. The boy's death had been almost more than Sylvia could stand. Plunged into a deep depression, she had finally required hospitalization for a time. Lately, though, taking care of the new baby they had adopted to help fill the vacuum, she seemed to be regaining her enthusiasm for life.

There were other problems. After twenty-five years underground, Albright's life in the coal mines was beginning to take its physical toll. Since 1963, he had been drawing disability benefits for the bad case of silicosis which made his breathing painful and difficult. Only a few weeks ago, the doctors had also determined that Albright had miners' pneumoconiosis—black lung—and had certified him for additional benefits. But he kept on working. He felt that he and Sylvia were "Living for them two boys, they was our whole life," and he had been glad enough to sacrifice his health for them.

Friday night's hoot owl shift had been colder than most, but otherwise routine. Working with his helper, a fellow-miner named Tunis Sipple, Albright had spent most of the shift repairing a malfunctioning coal loader. Halfway through the shift, he and Sipple had paused, fired up an old welder which they used to keep themselves warm, and eaten the lunches they carried in their tin pails. After the half-hour break, it was back to work. The wiring problem inside the coal loader was the kind of challenge Albright enjoyed. It was part of the reason he preferred coal mining to other jobs. Whenever he thought about doing some other kind of work, Robert remembered his ill-fated experiment of a dozen years ago, when he had left West Virginia to take a job in a Dunkirk, Indiana, bottle factory. The work was boring, endless ("Seemed like it took your shift forever to go by!"), and after only two months, Albright had repacked his bags and returned to Buffalo Creek. The coal mines might be gloomy and dangerous, but at least they offered a variety of jobs and challenges.

Now the belt rattled through the underground darkness. Albright lay on his back staring at the roof of the mine as he was pulled toward the surface. Idly, he wondered if his family would be waiting in Lorado to greet him. Steve and Sylvia had planned to drive up to a band concert at Morris Harvey College in Charleston that morning, but the heavy rains might have meant a last-minute postponement. Robert figured, as he rode the vibrating belt a few yards from Tunis Sipple, that they were probably still at home.

Suddenly, the belt stopped. For a moment, he lay motionless in the dark, half-expecting it to begin turning again. But nothing happened. As Albright and Sipple climbed off the belt and began a half-mile walk back to the portal, they discussed the sudden power shut-down: Robert had never seen it happen before. The failure was irritating, but Albright knew nothing of the problem at the Three Forks dam, and made no connection between the two. Emerging from the mine portal at a few minutes past eight, he jumped into his lightning-yellow Gremlin and began the ten-minute drive out of the mountains and down the hollow toward home.

Entering Buffalo Creek Hollow at Pardee, Albright was stunned at what lay before him. The hollow had become a funnel through which thirty feet of black water went plunging along, taking out every structure in its path. He gripped the steering wheel. One by one, the homes along

the road were being flattened. Already, the water had
torn the Pardee electrical station out of the ground—
hurling sparks and flame high in the air—and sending its
giant transformers bouncing down the hollow. Trapped
helplessly in his car, Albright prayed that his family had
gone to Charleston for the band concert after all. If they
were still sitting in the house at Lorado and had received
no advance warning, he knew they were doomed. Bolting
out of the Gremlin, he began to fight his way along the
rugged hillsides above the hollow. It was more than half
a mile down to his house at Lorado. Ignoring the pain in
his lungs, he battled through the thick, tangled scrub. But
his heart sank inside him. Already, he knew he was too
late.

The Albrights, as it turned out, had received no warning. Seventeen-year-old Steve was standing in the backyard when the water arrived. He raced back inside the
house, and a moment later emerged with his mother, who
carried nine-month-old Kerry Lee in her arms. Desperately, the three fought their way through the rising water,
almost reaching the hillside which stood a few dozen yards
behind their house. But they had started too late. The
water rose to their waists, then to their shoulders. The
current began to push them down the hollow.

Neighbors who survived the flood later described a
pathetic scene: Sylvia standing almost at the bottom of
the slope, swinging the baby back and forth through the
air, trying to find the strength to throw him up to the
crowd on the hill. Finally, the child dropped out of her
arms, and, in clear view of the horrified peeople above,
all three of them were swept away.

A few minutes later, reaching his own neighborhood at
last, Robert Albright had his worst fears confirmed. Everything was gone. Only the bare foundations remained to
show him where his house had been. The main wave had

passed through Lorado by now, but the water was still shoulder-deep in most places. Mindlessly, Albright plunged in. He would swim his way out to where his house had stood. Somehow, he would save them yet. Fighting his way across the torrent, he soon tired to the danger point.

Floundering, choking on the water and the coal-black sludge, he was close to drowning when some men standing on a nearby bank finally threw him a telephone wire. He pulled himself out of the flood on the rescue line, paused to recover some of his strength, and then began asking neighbors if any of them had seen his family. A few minutes later, his last hopes were dashed: the baby had been found wedged in a culvert, face-down, about 100 yards below the house. Had his family gone to Charleston after all, Robert knew they would first have dropped Kerry Lee off at his sister-in-law's house, out of the danger area.

Moving slow as in a dream, the numbed Albright limped to the nearby house where they had taken Kerry Lee. The little boy was in bad shape. "He was coal-black all over, he looked just like a tar baby. He had a whole patch of skin tore out of his head, and his leg was cut to pieces. They had been working on him—trying to get all that gob out of his throat."

But then a wonderful thing happened. The baby, which had not made a sound since his rescue, began wailing the moment Robert picked him up. It was a strong cry, and Kerry Lee kept it going. Albright figured the battered child would live—if only he could get him to a hospital in time.

Climbing into a neighbor's four-wheel-drive truck, with the baby wrapped tightly in a blanket, Albright began a four-hour nightmare. The only road out of the hollow was blocked by a rock-slide. Albright was forced to sit helplessly in the truck while they cleared the slide, even cutting several trees out of the way. Every once in a while, they had to get out and walk while the truck was pushed through rough spots. "We were stumbling along in all this mud and rock. I thought I was gonna break in two."

Finally, they reached the hospital and the baby was rushed to emergency. For three days, Robert did not once leave the child's side, did not even change out of his grimy work clothes until, on the third day, friends brought him a fresh set. Kerry Lee pulled through the crisis, and the story of his incredible escape quickly passed among the Buffalo Creek residents, who have referred to him as "The Miracle Baby" ever since.

His young son was out of danger, but the bodies of his wife and oldest son had been found, about 800 yards below the home in Lorado, and five days after the flood Albright would have to go down to the morgue to claim them: "My son was crushed up so bad, I went about four times trying to identify him. His head was just smashed to jelly. He had just a little bit of sideburn left, where you could tell it was him. All the bodies had swelled up so bad, you had to just keep looking and looking. . ."

Robert Albright had sacrificed much of his health to provide for the family he loved. Now, with the oldest son's death in Vietnam, all the members of that family except Kerry Lee were gone forever. "It was just like a whole lifetime went with a snap of a finger," Albright says. "I killed myself working up there in those lousy mines—but they only killed me little by little. Now that baby's all I got to live for. I tell you, if it wasn't for that child, I wouldn't be alive today."

Albright gave up on coal mining. Content to draw his $398 a month in disability payments, he can be found today in one of the temporary trailer parks along Buffalo Creek. He spends his days fixing the baby's bottle, changing the baby's diapers, and occasionally wondering, after all the years of work, what has happened to him.

In less than fifteen minutes, the killer flood had done its work in Lorado. Now the lucky ones, the people who had skirted death at the very last moment, gathered to tell each other of their heart-stopping escapes. J.C. Glend, for example, a sixty-nine-year-old retired coal miner, remembers that he was playing with one of his five small dogs in the bedroom when the water slammed into Lorado. He screamed for his wife Bertha, then in another room of the red-roofed house.

"I looked up and it was coming against the house," Glend said later. "It was like 9,000 cyclones out there. There was timbers and everything in it, and the water was above the windows. This was what you call a bust-out!

"Then I looked and here come the house next to us floating by. [It belonged to coal miner Florencio Sosa and his wife Magdalene: both drowned.] I can't hardly tell you how it was, it was so shocking. Then the water just lifted us up, and in five minutes we was up against the schoolhouse."

The Glends, still standing in their own home, had floated about fifty yards downstream, coming to rest against the walls of the Lorado school. Bertha was seriously injured, but both lived. "I knew if I tried to get out, that rock would beat me to death," Glend said of his wild ride. "You know, I'd say I've come close to getting killed fifteen or twenty times in the mines. I was used to danger. But I never saw anything like this—and I hope I never do again!"

Shirley Adkins's escape was even more remarkable. For Shirley, a slender, blue-eyed brunette, February 26 was to have been just another Saturday, with her husband David

putting in his regular shift as a coal loader at Buffalo Mine
No. 5, and she taking care of the couple's two small
children. David had worked a double shift on Friday.
Dead tired, he had dropped into bed by 9 o'clock Friday
night and slept soundlessly through to breakfast. Shirley
remembers the heavy rain drumming on the roof as she
fixed their usual morning meal—cream of wheat, toast,
coffee and Pepsi-Cola—and then called her husband to
the table.

The children—four-year-old David, Jr. and three-
month-old Dorinda—were still asleep. As they alternated
mouthfuls of creamed wheat with sips of Pepsi, the husband and wife discussed the coming day, and the flood
warnings they'd heard on the radio. Then David grabbed
his lunch pail, kissed Shirley good-bye, and departed for
No. 5, where his shift would begin at 7:15.

Surprisingly, he was back within an hour. Only a few
of the miners on his shift had showed up for work. David
figured that most of them, assuming that the roads would
be blocked by the heavy rain, had simply skipped their
shifts. And there was other, unsettling news. After listening to several scary stories about the threatened dam,
David and Shirley's father had decided to go and look at
it for themselves. It looked bad. The water had been rising, David said, and by now lapped close to the top of the
embankment. The company had several men at the scene,
and the officials were saying everything would be all
right.

The whole thing was a little unnerving, but then, David
and Shirley were safe in their own home now, and what
could happen there? David went off to take a bath, Shirley returned to the dishes, and five minutes later the
lights flickered off.

Almost at the same moment, there was a knock at the

door. It was David's sister, Glady Necessary, and she was screaming: "The dam's broke! Get out of here, the dam's broke!"

The Adkins had perhaps a minute in which they might have escaped. But they wasted it, while David roamed around the house looking for his son's glasses and Shirley tried to call a relative on nearby Huff Creek with news of the flood (the line was dead). Finally, Shirley grabbed the baby, David picked up his young son, and the four of them dashed out onto the front porch. They stopped, horrified. They were too late. As they watched, a churning wall of water lifted Glady's house, next door, out of its foundations and sent it spinning toward them. There was nowhere to run. "We just stood there for a second," Shirley remembers, "and we knew we were gone. Then I said, 'David, let's lay down here on top of these kids.' "

The family stumbled back into the living room and lay down on the floor. Shirley covered Dorinda with her body, and David lay atop his four-year-old son. The water arrived. The front porch disintegrated. "The wood tore off, you could see it going, and the noise was terrible with everything crashing." David told his wife, "Honey, we better start praying." The new wood-paneling bowed inward, splintering. A dish cabinet turned over with a crash. Helpless, they stared transfixed at the black water now rising through the new carpet. Inch by inch, the water crept up the walls. They climbed into the tallest chair they owned—Shirley sitting on the back of the chair with Dorinda; David kneeling on the seat with his son. The long silence was broken again as David said, "Don't worry, honey, we're in the creek now."

Almost hypnotized, Shirley stared deep into her husband's eyes. The water had reached his shoulders. She saw two tiny drops splatter against the side of his face. She heard the four-year-old whimper "Daddy, help me," and a moment later the water covered both of them up. "The next thing I knew," she says, "the house had caught me by the head and I was floating down the creek." Her head wedged painfully between two boards, Shirley was drifting downstream with the house breaking up as it went. Stunned, floating along in a dream, she watched the Lorado schoolhouse go by as though she were seeing it from a passing boat. The house was twisting and turning now, cracking to pieces around her. She struggled to keep a hold on little Dorinda. Half a mile farther, at Italy Bottom, something knocked her senseless and she lost the baby. "The pain was getting bad and I knew I was dead. I wanted to die, too. The others were gone—I wanted to go with them. I think I started to drown then. I know my head was down in the water, and I passed out."

Suddenly, she was conscious again. She had bobbed back to the surface in Lundale, in front of a row of two-story homes which housed the bosses from the nearby Amherst Coal Company. Covered with black sludge, she was loose in the water, in a vast ocean of wreckage. "Cross-ties or boards or something kept hitting my legs—I thought they was tore off. Without me even knowing it, my arm was over a crosstie and the debris was up so high, I just climbed up on it."

Shirley climbed aboard a floating mattress—so close it seemed to have been sent for the occasion—and began a strange ride down the creek. At one point, she noticed a little boy on an adjoining mattress, the two of them riding neck-and-neck, ducking their heads together as they passed under a railroad bridge. Now her fear was that she would float all the way down to Guyandotte River. She couldn't swim. What if the mattress sank in deep water? But a few minutes later, as she rounded a curve, the make-shift raft swung up against a hillside and stopped. Like a departing passenger, she simply stepped ashore.

Shirley ended her tumultuous ride at Stowe Bottom—almost four miles from where her house had stood in Lorado. They threw a blanket around her, and tried not to listen to her screams: "Get to high grounds!" she shouted over and over again, "Get to high grounds!" They found David Adkins later that afternoon, buried in a pile of wreckage. The body of David Jr., was not discovered until the following Wednesday. And to this day, little Dorinda has never been found.

Terrible as the deaths of David Adkins and his two children had been, they were at least swiftly accomplished. Such was not the case with thirty-seven-year-old Goldie Sipple of Lorado.

The wife of coal miner Tunis Sipple (Robert Albright's partner on the hoot owl shift at Buffalo Mine No. 8B), Goldie was washed several hundred yards downstream from her home when the water struck. She crawled from the wreckage more dead than alive, after inhaling great quantities of water and oily sludge. Neighbors took her to an undamaged home on a hillside above Lorado where they wrapped her in blankets and waited for the first rescue vehicles to arrive. But Goldie kept choking on the oily water she'd swallowed—choking and screaming in pain and then passing out again. For more than three hours, twenty people, including her husband, sat helpless in another room of the house, listening to Goldie Sipple die. "It was the most horrible thing I've ever been through in my life," a witness said later. "None of us knew what to do—we didn't know any first aid. She just kept gurgling and choking, she'd swallowed all this water. And she was screaming, 'God, why won't you let me die? I can't stand it anymore. . . . Why won't you let me die?'"

In the midst of so much anguish, there were still a few Lorado residents who could laugh, however. One of them was an elderly, retired coal miner named Roy Hicks, a tall, spindly, owl-faced man with a sense of humor. The Buffalo Creek flood startled him. Hicks admitted later, like he had never been startled before:

"I'm not ashamed to say where I was when the water hit. I was in the bathroom, sitting on the pot. I thought it was a train going by, but I thought it had a wreck. And all at once the door bursted open in my bedroom and before I could get my pants up, here I was with the water clear up to here [pointing to his knees].

"I finally waited 'till the water went down. When the water went down, I turned around, took off my clothes and went to bed, believe it or not!"

CHAPTER 5

Lundale

Pushing dozens of shattered houses before it, the flood-crest surged out of Lorado. The next community in its path was the little village of Craneco, where ten out of twenty homes were quickly demolished. A moment later, the wave crashed into Lundale, the largest town in upper Buffalo Creek Hollow, with a population of more than 700. The highest death toll of the disaster would be recorded here, as all but about twenty-five of Lundale's 150-odd homes were wiped out.

The flood triggered a thousand different rumors, predictions, and horror stories of one kind or another. Perhaps the first such unfounded report was the one which spread soon after the main wave had ripped through Lorado: another dam, even larger than the one which had just collapsed, was about to break. Many of those residents who had escaped the initial flood believed this prediction, and spent hours huddled in fear inside the undamaged houses to which they had fled for shelter. Children grew hysterical, screaming that another flood was on the way and they were about to drown. The parents had no way of checking on the rumor; they knew only that there were dozens of coal piles like the one which had just collapsed around Buffalo Creek Hollow. "That was the worst thing I ever lived through," remembers one housewife. "The kids were all crying, and everybody kept saying that another flood was about to get us. We didn't know what to do. We just sat there in fear."

One of the most popular stories born during these frantic hours of destruction described Steve Dasovich, the Buffalo Mining Company official who was responsible for the dam which broke, as having a nervous breakdown soon after watching the disaster unfold. Dasovich was supposed to have run through the hollow screaming, "Oh my God, I killed all them people! I killed 'em all!"

Dasovich did go into shock soon after meeting the flood at Craneco, and was hospitalized for several days because of it. He admits that his mind went blank, and "I don't remember too much after that." But no eyewitness has yet been found who quotes him as saying, "I killed all them people."

Another favorite, if apocryphal, tale was the one about the old woman who sat quietly, unmoving, in a rocking chair while her house smashed to pieces around her. "The Lord told me not to move," the old woman was supposed to have remarked later, "so I didn't. The Lord told me I was going to be all right, and I believed him. I just sat in my living room, praying for the others." Of course, the pious lady escaped without a scratch. . . . Then there was the recurring story of "This woman up the hollow, who left her kids to die in the flood." In that one, an angry woman who had been deserted by her husband, and who resented having to care for several small children, simply locked her doors and left the kids to perish inside the house as the wave roared through.

Not all of these stories were apocryphal, however. Michael Parrish, a twenty-year-old salesman who had grown up in Lorado and then moved to Detroit, got the best news of his life in the days which followed the disaster. After being notified by the Red Cross that his mother had died in Lorado, Parrish called three different Logan County funeral homes in a futile effort to locate her body. Frustrated, he drove at breakneck speed ("I almost had two accidents getting down here!") to Buffalo Creek Hollow. At first, the National Guardsmen refused to let him enter the devastated hollow. Parrish finally reached the family home in Lorado and found all of his relatives safe and sound. The Red Cross had been mistaken. "I walked in that house over there," Michael said later, "and there they were. All I could do was cry and grab my mother and hug her."

There were pathetic cases as well—like Shirley Adkins of Lorado. When the authorities were unable to find the body of her baby daughter, Dorinda, Shirley refused to believe that the child had been killed in the flood. "She went around for days," a neighbor recalled, "telling people that strangers had kidnapped her baby and taken it over into Kentucky. For awhile, we were afraid that she was losing her mind, but she finally snapped out of it."

More than one legend was born in the Buffalo Creek disaster. Within a few weeks, in fact, a Lundale flood victim named John Bailey was well on his way to becoming a folk-hero. Like "Steel-drivin' John Henry" (also a West Virginian), Big John Bailey was described as a man who had dared to do battle with the huge, heartless company—and had been killed for his efforts. "Big John Bailey knew it was going to happen," said Don Bryant, chairman of the militant Logan County Black Lung Association, in a published statement. "Big John Bailey was one of the strongest members of the Black Lung Association in Logan County. He lived in Lundale with his wife and step-daughter. For the last years of his life, Big John told the news media and people throughout West Virginia about the slag dam at the head of Buffalo Creek. He warned that someday there would be a disaster in this hollow. Big John, his wife, and step-daughter were buried in the mud when the dam at the head of Buffalo Creek broke. Nobody listened to his warnings and his reward was death." There was little evidence to support this martyr-like description of Bailey, but no doubt at all that he was rapidly becoming a symbol for angry miners who saw the disaster as another example of the economic oppression visited upon them by the coal companies.

Sound asleep in his Lundale home, a young coal miner named Larry Owens would be rudely awakened this morning, as the floodwave ripped through town. . . .

"Bubby, bubby, get out of bed quick!" Larry's sister Berma Jo danced back and forth at the foot of his bed. "Oh bubby, the dam's broke!"

Larry—they always called him Bubby at home— wrapped a blanket around himself and rolled to his feet. Standing before the downstairs bedroom window at his parents' home in upper Lundale, Owens looked and caught his breath. Berma Jo was right—the big Three Forks dam must have collapsed because six feet of water now lapped at the downstairs windows, and the house shook with a strange palsy, as though it were about to rattle apart.

But Owens—a sandy-haired, twenty-two-year-old coal miner—wasn't one to panic. It wasn't in his character. Easy-going and relaxed, the chunky Owens had grown accustomed to danger in the deep Amherst Coal Company mine where he worked. And he had already experienced two minor floods in the past. The water looked bad, but it didn't frighten him yet. Everything would be all right, if they just kept their heads.

From an early age, Owens had learned to take care of himself in an unpredictable, often hazardous world. The son of a career coal miner, Larry and his three brothers and sisters had weathered the inevitable accidents that went with their father's profession: the broken neck he had received in a buggy accident, the broken pelvis he had suffered in a sudden slate fall. These injuries had not stopped Onsby Owens. He went right back to work—and right back to teaching his kids the simple, fatalistic philosophy he lived by: "If it's your time to go," Onsby always said, "then you're going to go. You might as well stop worrying about it."

Larry had improved on the philosophy, adding a dash of independence, an almost bull-headed determination to go his own way. They told him he was too small (at 150 lbs.) to play tackle on the high school football team: he made all-county tackle. They told him he was too young (at seventeen) to get married: he married Henrietta Cook anyway, and the marriage prospered. They told him he should live along Buffalo Creek and work in the mines like his father: he packed up his young family and moved to Indianapolis, to another line of work.

But the most recent experiment had been a little disappointing. For four years, Owens had labored in an Indianapolis garage to support Henrietta and the two baby boys. Gradually, he and his family had grown weary of their big-city life: "It was one big rat race, with the traffic and all. Everybody's going everywhere, and nobody's going anywhere. . . . Finally, I realized that I loved the hills of West Virginia."

Only a few weeks before, then, Larry Owens had come home again to Buffalo Creek. It had not taken him long to land a job with Amherst, where he now drove a coal buggy on the afternoon shift. Owens had little trouble adjusting to the job—he'd already put in a year underground before the move to Indiana. Soon his life returned to its old patterns: the long work-hours spent in the dusty gloom loading coal, the days off spent hunting and fishing in the mountains around the hollow. Now and then, there was a night out with Henrietta: playing cards at some friend's house, usually. And once in a while, Larry returned to

The Junction Drive-in, a favorite hang-out in Man, where the miners sat guzzling beer and talking about what had happened on the latest shift.

It was one hell of a way to make a living, Larry admitted that. For one thing, coal mining meant existing with danger. You simply had to accept it, stop fighting it. The real test came during the first few weeks. If you could handle the fear during this breaking-in period, then you would soon be able to ignore it. Like most of the others, Owens had been frightened at first. "When you first go down into those mines, you're scared. You ride down there in a mantrip (mine car), and they tell you to keep your head down. Well, my head was down lower than anybody else's!"

Some men never got over their fear. It showed in their faces, in the way they worked: "Say a piece of rock kinda dribbles from the top . . . and the guy next to you kinda takes off running. You say, 'Where you going, man?' He says, 'I'm getting away from there, man, that thing's about to fall!' " After a few weeks, however, the average miner will be able to tell everything by sound. A creaking roof, a "dribble" of rock might be perfectly harmless. Then again, one short cracking sound could signal a deadly fall, perhaps only a few seconds away. You had to keep thinking.

Once adjusted, and secure in the knowledge of the big union behind them, many miners turned their attention to getting out of work. It was no secret—the work pace was usually a kind of compromise hammered out between the foreman and the workers, who would not let themselves be pushed too far. Sometimes, the ploys used to avoid work were surprisingly devious: "Let's say the boss tells you to go get something, like a bag of rock dust. Well, you can be gone half the night, and he can't say nothing to you. You come back, he asks what happened, you say,

'Oh, man, I broke down on the way!' It's a lie, I know. But I've done it before. I'm not perfect."

Larry goofed off once in awhile, it was true, but one thing he never did inside a coal mine was sleep. "That's one of my rules," he says. "I don't ever sleep on the job. If I'm gonna die on the job, I want to see it happen. I don't want to get killed while I'm asleep."

By February of 1972, Owens was making $41.05 a shift. And although "You never have enough money with a family of four," his life had settled down into a comfortable routine. A few more weeks, and he hoped to move out of his father's house in Lundale to a place of his own. He knew that his wife Hen and his two small sons, three-year-old Michael and five-year-old David, would be happier with more room. Onsby Owens's home was just too crowded with all of them living there. . . .

Larry stood at the window trying to decide what to do. The others were up by now. Larry heard Anita (his younger sister) shout to her husband, "Lord, Henry, look at your car!" as the couple's new Ford Mustang went tumbling down the hollow in front of the wave. A moment later, Larry had decided on a course of action. He led the way back through the kitchen to the stairs, and all of them climbed up to the second floor. Crowded in an upstairs bedroom, Owens stood next to little Michael, looking out over the front yard. Houses were stacked up in the black water like cards in a huge, exploded deck.

Suddenly, right next to them, the bedroom wall shattered open and the roof of an adjoining house came barrelling in. The water had driven a floating home against their own. Larry stood in shock for a second, gaping at the shingles beside him. Then he picked Michael up and threw him headfirst into the next room.

Now, the floodwave was "crunching and cracking, like

thunder," and one by one, the windows in the Owens home were breaking out of their frames. "I wasn't scared when I saw the water," Larry says, "because I knew I had been through a couple of floods before. It's just water. But when that house came through there, I really got scared. I knew this was a bad one, and we might not get out."

They stood in the adjoining room, white-faced, not speaking. Slowly, the house toppled over "just like a tree." They fell sideways through the room, in a tangle of arms, legs and screaming children. Lying on its side, the house began floating down Buffalo Creek. If they were going to get out alive, somebody had to make a move. Cautiously, Larry climbed through one of the broken windows and balanced on the ledge. Suddenly it looked like they had a chance. For the time being, anyway, the house had lodged against something, and so stood motionless. If they jumped from the window ledge to the next roof, they would probably be safe because the adjoining house, standing partially out of the water, had withstood the flood.

Larry jumped first, landing easily on the nearby roof, and turned to help the others. Henrietta had already climbed out onto the window ledge above him. He held out his arms to her, preparing to break her fall. In that instant, the Owens house broke loose again and resumed its ride down the hollow.

They looked at each other. Henrietta held her arms out, also, only a few feet away. In an agony of indecision, she started to jump, then held back. The children. She couldn't leave Michael and David behind. They looked at each other. Foot by foot, the distance between them lengthened. Arms still outstretched, Henrietta yelled "Honey!" That was the last word she said, and the last time Larry saw her.

Frantic, Owens put his fist through a window of the

house on which he was standing, slammed the window open, and ran downstairs. But he was helpless now; his family was gone.

After a few hundred yards, the Owens home cracked to pieces and all of its occupants were thrown into the water. Bill, Larry's fifteen-year-old brother, managed to climb on the roof for a wild ride down the hollow. He caught a glimpse of Anita and three-year-old Michael, clinging desperately to a log. Then they disappeared beneath the surface. Later, Bill jumped from the roof to a tree branch and escaped.

Most of the others didn't. Both of Larry's sisters—Berma Jo and Anita—drowned. His son, Michael, perished along with Berma Jo's one-year-old baby. And Henrietta Owens, Larry's wife of five years, was killed with the rest. The survivors included Larry, his five-year-old son David, his brother Bill and Anita's husband, Henry Smith. The others would be found days later, buried in the wreckage.

"I was mad at myself," Larry recalls, "because I'll believe to the day I die that if I'd stayed in that room, I could have saved somebody. I wasn't getting out because I was chicken . . . but because I wanted to help them out. If it [the house] had just stayed there thirty or forty seconds more, I'd have gotten them all out."

There was one bright spot—the discovery that David was still alive. "I just rubbed him and cried," Larry remembers. "I was so glad to have him back." But soon a neighbor told him that they had brought Henrietta into the morgue. "She was scratched in the face a little, but that was all," Larry said. "Berma Jo looked like somebody had beaten her in the face with a hammer, and her shoulder was separated from her body. Hen looked just like herself, except for—what's the stuff that sets in and hardens your bones? Rigor mortis? Yeah, that's it."

Now a long period of depression began for Larry

Owens. Three times in a row, he went to the mine for work and then returned home, unable to do his job. "I couldn't find any sense in working," he says. "I didn't have anything to work for. I'd just come home and watch TV and think about the thing—think about how it would have been different if I'd tried something else."

The Buffalo Creek flood virtually annihilated Lundale, West Virginia. At least fifty of the town's 700-plus residents were killed; most of its homes were washed miles down the hollow. Both the Lundale Free Will Baptist Church and the Island Creek Supermarket suffered heavy damage. Once again, however, a major coal company installation escaped destruction: Amherst Coal's Lundale headquarters were almost untouched by the racing wave, and were soon back in operation.

Most Buffalo Creek residents were completely unaware of the danger they faced on this stormy Saturday morning. For many of them, only a few minutes' warning could have meant the difference between life and death. . . .

At Three Forks, for example, coal miner Jason Bailey, ignoring the advice of his neighbors, had departed for work at 7 A.M. that Saturday. An hour later, while Bailey struggled to make his way back to the family's devastated house, his wife and four of his five children perished. Denny Gibson describes the moment when Jason discovered his wife's body:

> I looked over on the side of the mountain, and at first I thought it was a big doll. That's what I wanted it to be. I said, "Lord God, have mercy." We walked across and Jason went with us. And there was his wife, lying within three feet of where we'd found that little boy (earlier).
>
> She didn't have a stitch of clothes on her. They was completely cut off from her. Jason reached out and got

ahold of her hand. I said, "Lord God, have mercy." He walked up and took ahold of her and tried to cry and it seemed like he couldn't. He was shocked so bad. Even myself, I can't explain just what kind of feelings I had. It was a nightmare. I went through the Korean War, and I never saw anything like this. . . . Never.

Mearl Bartram's anguish was similar. A clerk for the Amherst Coal Company, Mearl lived near the Lundale Supermarket with his wife, Joyce, and her parents, Milton and Effie Baker. Knowing nothing of the danger at Three Forks, Mearl left for work at Amherst's nearby Rum Creek office around 7 A.M. He telephoned his wife at ten minutes to eight—to assure her that there was no flood danger. The creek was up a little, he said, but nothing to worry about.

At 9:30, the power in Mearl's office suddenly shut down. By then, the Logan radio stations were broadcasting increasingly ominous bulletins about a flood along Buffalo Creek. Mearl hurried back to the hollow, and gasped as he saw the first of the debris. There was nothing he could do. "I couldn't get up there [to their home in Lundale], and it was hopeless to think about looking in all that wreckage. So all I could do was wait."

It was a long wait. As the hours passed, and Joyce failed to call in on one of the Amherst mobile radio trucks, Mearl lost hope. A truck driver came by to tell him that the Bartram house had vanished without a trace. It was a week before Mearl, walking among the bodies at the temporary morgue, identified his wife. Her parents had also perished.

Later, Mearl moved into a tiny room at the Logan YMCA. "I had a home," he remembers, "that when you left in the morning, you was kissed, you was cared for. Your breakfast was got for ye . . . if it was raining, they cared whether you had something on your head when you went out the door. At supper, we'd have a good hot meal, and then read the Bible in the front room. They was all concerned about me. I never had anything like it, the love they had for me. All this happened in the twinkling of an eye. Now I have nothing, no one that cares. You know what I mean—to stand by you if you're sick at night. You come down to a life where you've lost all your clothes, all your belongings, all your pictures of the ones you loved, everything. Down to a life. . . . I just don't know how it's going to turn out."

In Lundale, a few lucky ones learned of the coming flood purely by accident. Ruth Tomblin and her family were among these fortunate few. They were visiting Ruth's sister in Lundale on Friday night, and at one point, the sister decided to listen in on her telephone party line ("Just curious, mind you!"). The conversation included no idle gossip: a frightened housewife was telling her absent husband, "The big dam has cracked, they say it's going to break."

"I'm petrified of water," Ruth Tomblin said later, "so I practically made my husband take us all down to my sister's house at Man." Ruth and her several small children were miles from the danger area when the water struck.

Others escaped at the last moment, when they heard shouted warnings or happened to sight the flood in the distance. Cleo Collins, a fifty-two-year-old widow, was serving fried eggs, bacon, biscuits, and gravy to her family in Lundale that Saturday morning. ("I never did get the gravy!" her daughter Jo Anne remarked wryly later, remembering the moment when the water arrived.)

While the family sat at the breakfast table, a neighbor pulled up in the yard outside. "He came by and blew his horn and hollered: 'The goddam dam's broke, get out of here!' I grabbed Momma's shoes and her pills (for a heart condition) and we took off. We had about three minutes. As we crossed the tracks, the water took our house. And when I looked again, I saw another house going by with people in it. After that, I didn't want to see any more."

Leaving Lundale, the floodwave shot downhill toward Stowe, a tiny community of about thirty houses. More than five miles from its source, the wall of water was still more than ten feet deep in places. Sound asleep, many of those in Stowe Bottom never heard it coming.

Stowe Bottom

Only moments from the disaster which would alter his life forever, Wayne Brady Hatfield slept on, his broad, craggy, fifty-eight-year-old face still closed to the new day. Wayne Hatfield: direct descendant of "Devil Anse" and all his battling clan; six feet, five inches tall; 280 pounds of jutting shoulder and mulish back. Ponderous afoot, a deep-fisted, scowling hill of a man.

He belonged more to the past than the present, to the flashing creeks and the deep, sun-checkered hollows of his boyhood, to moonshine whiskey in a stone-chip jug and blue-black hogs snuffling in the August mud. His roots went to Hatfield-McCoy, at the turn of the century—to the hard-eyed men who glare at you, angry in their beards, from yellowed photographs. He came out of mountain songs like "Rattler Was a Fine Old Dog" and "The Wildwood Flower." More so than most, he was a link between the old, hillbilly world of West Virginia and the new, changing West Virginia of color television and the flameless electric range.

Wayne Hatfield grew up with thirteen brothers and sisters—the children of Leander ("Big Lee") Hatfield—on a farm along rural Horsepin Creek. His parents' world had already begun to change by then: the bitter Hatfield-McCoy feud fading into memory, the new roads and the new radio opening West Virginia up to the rest of the country. But the old ways lingered on. It was a childhood built on his father's stern, God-fearing religion—and on

Big Lee's talent for making moonshine whiskey. "My daddy was a farmer," Hatfield says, and then with a sideways, secret grin, "We farmed . . . and, well . . . I guess I just might as well say the fact. We made moonshine whiskey and that's all we ever lived on!"

The Hatfields grew their food, and moonshine provided the money for other things. "We lived twice as good as I've got today," Wayne says. "Why, they say that back in Hoover's time, everybody starved. I don't know whether they did or not—but I know you could buy a sack of potatoes for sixty-five cents. Why, right up there at Gilbert, I've bought a hundred pound of taters for sixty-five cents. . . ."

Wayne and his brothers and sisters divided the chores that went with running a farm, and then pitched in to help with the moonshine. Hatfield remembers building the still, moving in the mash barrels, setting up the "thumping keg" with its condensation coil, or "worm," and then watching while the "corn chop" or fermented peaches filtered drop by drop into mountain liquor. "It didn't take much of it to get you! And that was good whiskey; you didn't get no hangover unless you stayed drunk for three or four days!"

Once in awhile, the "Revenooers" or the state police came around, but they were easy to fool. "I seen half a dozen state police come up one time and go around the back side of the field. They didn't find nothing—and I'd say there was a sack of moonshine in every fence corner!"

Sometimes the moonshine business produced hilarious complications. One of Hatfield's five sisters, Mrs. Russell Harry, remembers an evening when she and her brother were sent to bring home the cows. On the way out to the pasture, they stumbled accidentally over two of Big Lee's mash barrels. "There was apricots and peaches in 'em," Mrs. Harry says, "and we sat down and started eating the peaches out of that, and that mash made us so drunk we couldn't get back to the house! We didn't go hunt the cows at all. And when Mommy come up and found us, she really sobered us up good. Boy, she wore us out!"

By 1933, Wayne Hatfield was nineteen years old and looking for work. Like thousands of other young West Virginians during that era, he wound up inside a coal mine—as a conveyor belt operator for the nearby Mallory Coal Company. It was hard work, but the massive, block-shouldered Hatfield took it in stride. "I was stout as a horse," he says of those early days. "Why, I don't figure there was five other men could have done anything with me. I was stout!" Hatfield enjoyed feats of strength. Let the other men pull their conveyor belt sections through the mine with ropes; Hatfield dragged his section along single-handedly. Let two or three of the miners labor to remove a heavy engine part; Hatfield would lift it out with one hand.

The work was dangerous, as well as tiring, but Hatfield refused to be intimidated. He watched his share of accidents over the years. "I've seen rock fall on 'em, mash 'em all to pieces. I seen Ben Dean, a man I worked with back then, a rock about two feet across, caught his head on the top of the loader and his light went through his head. It busted his head open. I brought that man outside.

"He didn't die. But just as sure as you and me are here, some of that man's brains run out through his eyes. After I come out of the mine that day, the lamp man asked me, said, 'Hatfield, Ben's brains was running out his eyes, wasn't they?' I said, 'If I aint the biggest fool on earth, they was. . . .'"

Hatfield's first bad accident came in 1944, when a 500-pound "kettlebottom" (a solid lump of rock) fell from the mine roof and struck him. The crushing blow injured his spine, cut the nerves and tendons in his hip, and left him

with the miners' most dreaded affliction: a bad back. The
mine bosses told Hatfield he would never work again, but
he surprised them. After a few months of recuperation, he
went back on the job and stayed on it until his disability
retirement in 1969. There were other injuries: a broken
hand at one point, a fractured foot at another. But Wayne
stuck it out. He wasn't quite as mobile after the back
injury, but he was strong as ever.

Usually a quiet, soft-spoken man, Hatfield became
furious whenever he felt he'd been cheated or lied to. In
one celebrated case, a local Chevrolet salesman made the
mistake of selling him a lemon. "I traded cars with 'em,"
Hatfield glowers, "and that salesman gave me one of those
old cars—it wasn't worth nothing. I had to go back and
whup that salesman. I whupped him upstairs and downstairs. I just tore him up like a turning plow turning new
ground! I didn't say nothing to him, I just walked up and
knocked the shit out of him.

"They tried to get me off him. Ralph Hatfield started
throwing things at me from out of the parts department.
They was throwing parts at me, but they couldn't do
nothing with me. So finally, they all run off and left me.
I guess you could say I owned that Chevy garage for
about fifteen minutes! Yes sir, that was something. That
boy went down them steps just like a whirligig!"

After marrying Etta Pearl in 1940, Hatfield had fathered
seven children, providing for his big family by working all
the hours of overtime he could get. He taught his kids the
same God-fearing, down-to-earth philosophy Big Lee had
taught him: a man's word was his bond; the worst sin you
could commit was to tell a lie. As the years passed, the
children grew up and moved away to raise families of
their own. By 1972, Hatfield had been retired for several
years. Drawing his $210-a-month Social Security and his
$150 miners' pension, he lived with Etta Pearl in a two-

story, seven-room house in Stowe Bottom, a community of twenty-one homes located just below Lundale. The green-roofed, $4,000 home was pleasant enough, with its three elm trees in the yard and its separate garage. It was a quiet, relaxed life. Wayne's bad back and his crippling case of black lung prevented a lot of activity. Mostly, they sat at home. "I'd piddle around picking the guitar and reading the Bible most of the day." The guitar was Hatfield's favorite pastime. He spent hours practicing the old songs he'd picked up from A. P. Carter and the Carter family, a hillbilly group he especially liked. The drowsy summer afternoons found him sitting on the porch, singing in a squeaky, breathless voice, tunes like "Rattler Was A Fine Old Dog":

> Rattler was a fine old dog,
> As fine as he could be.
> Every night at suppertime,
> They'd pay that dog a fee!

Or his own, rumbling version of "Little Brown Jug":

> If I had a cow that'd give such milk,
> I'd clothe her in the finest silk!
> I'd feed her on the choicest hay,
> And milk her forty times a day!
> Hoo-haa-haa, it's you and me,
> Little brown jug, do I love thee!

And so the days passed at Stowe Bottom, picking the guitar, reading the Bible, watching television now and then. Etta Pearl was a good cook who often fixed Wayne's favorite treat, "soup beans," and who made mealtime something to look forward to. These days, the Hatfields entertained two extra guests at every meal: their daughter, Judith Annette, was recently divorced, and lived with her own baby daughter, Connie Sue Ferguson, at the Hatfield place in Stowe. But the four of them got on well together, and Wayne enjoyed the baby's antics more every day. It was a relaxed, rewarding existence, one which Wayne had worked all his life to enjoy.

Now Hatfield opened one eye, and scowled at the ceiling. Something wasn't right. Etta Pearl snored quietly beside him; Judith Annette and little Connie Sue slept soundlessly in another room. It was Saturday morning, the 26th of February, with pale, winter light falling through the window and a patter of rain. But something wasn't right. That sound . . . a kind of rushing, rustling sound, like moving water . . .

He sat up. He saw it. The Hatfield place was completely surrounded by water. They were caught in some kind of flood, and there was no way out. Hatfield hurtled into the living room, shouting, "Lord, have mercy!" and then turning to yell up the stairs at Judith Annette: "Get that baby down here as fast as you can! We're caught in the water!" He looked again, and a blast of fear hit him: the garage had floated right out of the ground, taking his blue Ford truck with it, and was now gliding down the hollow.

Judy flew down the stairs; Wayne grabbed the baby and plunged through the front door out into the water. Gasping at the shock, he fought his way several yards out, then gave up. "I tried to run through the water," he said, "but it was too deep and I had to run back inside. It scared me to death. All I could do was stand in the door and pray and ask God to have mercy on us. 'Cause I knowed they was gone. I knowed they was gonna drown."

Ordering his family back upstairs, Hatfield remained in the doorway as the house began to break up. The windows were shattering out by now, and the beams in the ceiling cracking in half. He could hear them screaming upstairs; the sound buried itself forever in his memory: "They was yelling for me to come upstairs with 'em. They was scared

I was gonna drown. But I wouldn't move. I told 'em to pray just as hard as they could, and I'd try to watch a way out for us."

There was no way out. Soon the house broke loose from its foundations and began bumping down the hollow, cracking up as it went. "That was the awfulest rumbling and roaring that ever you heard on earth. There was cars, houses, saps, logs, that old mud and stuff—anything you can name, it was in there. That house was floating up as high as them telephone poles."

Suddenly the house slammed into a large tree. The walls came crunching together, and Hatfield blacked out. When he woke up, a few seconds later, he was floating face-down in the living room. Swimming now, he kicked his way through the front door. A sudden surge sucked him back in. Again he fought his way outside. Tiring now, gasping from his case of black lung, Hatfield was failing when his hand "suddenly clamped down on the railroad tracks." He pulled himself up on the railroad embankment, and was safe.

Hatfield had ridden the flood almost a mile. His mouth full of oily gob, he sat shivering on the tracks, praying that Etta Pearl and the others had somehow escaped. Finally he got up, and limping along on a broken foot, made his way down the line. After a few minutes, he found some neighbors who told him there were two bodies lying out in the open at Latrobe, about a mile on. Hatfield limped down.

"I found my wife and daughter down in Latrobe," he remembers, "lying in a pile of slats and stuff." Both were dead. Hatfield asked some people to help him carry the bodies up to a nearby house; the people were afraid to touch the bodies. "I begged and cried and prayed," Wayne says, "I done everything that could be thought of to get them in the house. But they wouldn't help me." It was

7 o'clock that night before the National Guard evacuated the bodies (Connie Sue's was found later), and Hatfield could leave for the hospital. He had survived the flood, but lost the best part of his life, and now the months of anguish began. He did not try to hide his bitterness: "They [the company officials] killed them people," he said, "just the same as if they'd shot 'em. It was pure murder, that's all it was."

For Hatfield, the disaster was only another example of the way the world had changed since his boyhood—the way people had stopped caring about each other:

> This country's changed, all right. America has gotten to be the low-downdest nation on earth. It's about as wicked as anything on earth. They don't care whether one or another lives, or whether they die. All they care about is making another dollar. Why, these big companies can just kill whoever they please, that's all, and that's the way she is, buddy.
>
> Back when I was a boy, growing up, say on a farm, if a man got sick, the folks would come down off of Bend Creek, off Horsepin Creek and Gilman Creek, why, they'd hoe that man's corn, take care of that sick man's farm for him. Somebody would go and see that he had coal carried in to build him a fire.
>
> If a man gets sick now—why, let him die! See, it's in the leaders of our country. The leaders don't care about the people anymore. America will holler: "Look what's going on over there, them little children starving to death!" They won't look around here, that they got little children starving to death right here in America. They take our money and send it over yonder to them people over there that we have to fight—and send our boys over there to get 'em killed for nothing. Just to make the rich man a dollar or two more. That's the way she is, buddy, that's the way they do.

Like many coal miners in Appalachia, Hatfield sees government as a mysterious, silent conspiracy between the wealthy and the politicians: "The rich men buy the politicians off, that's all. Why, I'd say right now, they've done already picked the next president of the United States. They've already set it out. They know just as much right now who's gonna be the next president as if he was already elected, I'll tell you that, I ain't scared. They might put me in jail, but I'm supposed to have freedom of speech. If we'd have elected George Wallace last time, that would have broke this other stuff up. Because George Wallace would have done what he said, or died a-trying.

"See, it would have broke this other stuff up. This government stuff ain't nothing but pure shit, buddy. God will forgive me if I ever write my name on another ballot. I'll never do it—I'll never vote no more. Never on earth, as long as I live. The rich man runs it all, that's the way I feel. The poor man ain't got a prayer in hell. Why, if a rich man does something to you, and you hit him back, the first thing you know, he's got you tied up in court so bad to where you got to let him go. It don't matter how much you want to whip his ass, there's enough law, that if he gets the law into it, he'll mess you up. If that ain't the way things are, buddy, then my name is McCoy! And I *ain't* no McCoy.

"I'm a Hatfield!"

By 9 A.M., the first reports of disaster had begun to trickle out of Buffalo Creek Hollow, as stunned residents telephoned the two Logan radio stations (WVOW and WLOG) to describe the flood. Soon both stations were interrupting their programs every few minutes, with flood bulletins that sounded more and more ominous. The reports were extremely difficult to check. Charlie Hylton, editor of Logan's daily newspaper, The Banner, spent the afternoon dialing Buffalo Creek phone numbers at random. After more than a dozen such calls either went unanswered or met a busy signal, Hylton decided to move, and dispatched a team of reporters to the hollow.

Before long, the authorities in Logan County knew they had a major disaster on their hands. Perhaps a dozen bodies had been recovered by noon, and the sheriff's deputies were phoning in incredible reports of wholesale wreckage up and down the hollow. The sheriff finally reached the National Guard headquarters in Charleston with a request for emergency assistance, and the first of more than 750 Guardsmen (who would spend two weeks working in the hollow) were dispatched to Man. Soon the Red Cross and Salvation Army mobile units would be rolling toward Buffalo Creek, to bring food, hot coffee, and desperately-needed medical supplies into the disaster area.

Meanwhile, as the still-lethal floodwave smashed through Stowe and Latrobe and Braeholm on its way to the Guyandotte River at Man, the stories of tragic—and unnecessary—death grew more and more numerous. At Latrobe (just below Stowe), for example, forty-nine-year-old Otis Ramey, the principal at the Lorado school, died when he attempted to return to his flooded house. After first fleeing to safety, Ramey had raced back to his threatened home to save the family dog. "He was standing in the door, yelling at the dog," a neighbor reported later, "and all of a sudden a pole came down and hit him in the back of the neck. It must have knocked him out. The water was getting deeper all the time. But his wife Mattie Ramey (also a teacher at the Lorado school) ran back to help him. She tried to pull him inside, and she was standing in the doorway when the house washed away." Both of the Rameys were killed.

A few doors away, forty-four-year-old Mary Marcum

ignored two last-minute warnings, and then perished. At
home alone with her seven pet dogs, she could not bear
to leave them in danger. "I sent my son Jimmy over there
to knock on her door," said a neighbor, Mrs. Ira Vincell,
"but she just waited too long and the water got her."
Six of the seven dogs died along with Mary Marcum.

Perhaps the most frightening, and dramatic escape story
to emerge from the Buffalo Creek flood was the one which
began around 9 that Saturday morning, when a panicky
Elmer Adkins rushed into his three-bedroom home at
Stowe Bottom and told his wife the dam had gone.

A slate picker at the nearby Island Creek tipple, and
the father of four children, Adkins had just returned from
his mother-in-law's house at Davy Hollow. The power had
shut down all over Lorado, Elmer said, and up and down
the road, frightened people were yelling that the dam had
broken, better run for high ground. Barbara Adkins was
hardly panicked. A plump, slow-moving woman, she had
been through all of this before. Was it 1967—when they
passed the word that the dam had collapsed, and made
everybody run? Barbara had dressed the kids (there were
only two of them, back then), and all of them had stood
around, waiting. Nothing happened. There had been some
flooding in Three Forks, if Barbara remembered right, but
at Stowe Bottom the creek just rose a little. Still, you had
to be prepared. With a sigh, she began rounding up the
kids: ten-year-old Jimmy Darrell, seven-year-old Theresa,
three-year-old Michael and the baby, Kathy.

After the children were warmly dressed, Barbara re-
treated to the bathroom to wash her face—and then began
applying a leisurely, full course of makeup. Elmer fid-
geted, stamping back and forth across the living room
floor. "Hurry up, honey! Will you come on? We gotta get
out of here!" Elmer's plan was to drive the family up to
the mother-in-law's house in Davy Hollow, which was

safely elevated above the Buffalo Creek floor. This plan failed.

("My husband said later," Barbara laughed, "that this was one time he was glad I was late! Because, as it turned out, if we'd have tried to make that drive up the hollow, we'd have drowned for sure.")

While Barbara applied the last touches of makeup, Elmer wandered outside to take another look at the creek. Suddenly he noticed several men, who stood on a hillside next to the Island Creek tipple above Stowe Bottom, jumping and waving their arms. What were they doing? They seemed to be pointing up the hollow, toward the next bend. They were shouting something . . .

All at once, Elmer understood. Standing high above the valley floor, the men near the tipple could see part of the way around the bend. They must have seen the water, the flood, racing toward them. Elmer turned, and sprinted back into the house. He grabbed the two smallest children, one under each arm, and headed for the back door.

"When we opened that back door up, it was just like a big river coming toward you. So we turned around and headed out the front way." The first thing the Adkins saw, as they opened the front door, was a neighbor's home being ripped out of the ground. One of the children shouted, "Oh Lord, Mom, there goes Fanny's house!" A moment later, a second home broke loose, carrying six people (the Waugh family) to their deaths.

The Adkins stood on their front porch. The water rose. One by one, the Adkins's cars began to float away. First, Elmer's mother's gold Rambler drifted down the hollow. Then a truck which had been parked in the yard glided away. Next, the Adkins's own 1965 Pontiac sailed off into the distance. They watched numbly while the cars departed, almost as if they were being driven down the hollow. "Elmer was in shock," Barbara remembers. "He

kept telling us, 'Now, when the Chevrolet comes by [it
was parked in the garage], I'll put you all in it.' I guess
he thought it was going to stop for us, or something."

The water kept climbing. Suddenly Elmer had another
idea: the pear tree which stood next to the porch. They
would climb high up in the tree, above the water, and if
it just held. . . . Jimmy Darrell went first, scrambling up
the trunk with the water swirling around his legs. "Jimmy
was scared sick," Barbara says, "he just kept climbing
higher and higher and his father had to talk hateful to
him to get him to stop. Elmer was scared the boy would
get up in them thinner branches and they'd break off."

Next, seven-year-old Theresa began the climb, followed
by Barbara and Ethel (Elmer's mother), who carried the
two small children. "It looked impossible," Ethel says, because the limbs were skinny and the trunk wasn't no big-
ger than eight or ten inches. But it held."

Elmer stood on the roof, next to the tree, and they
waited. If the house broke up, they knew it would probably carry the pear tree away with it. "I told Jimmy,"
Barbara remembers, "'Jimmy, get Theresa's hand and
hang on for dear life!'" The water inched up over their
feet, then their knees. They clung to the limbs, three-year-
old Michael repeating again and again: "Jesus, Jesus help
us."

They watched a body roll into the garage. A little boy
washed up against the cyclone fence and hung there.
"Elmer was crying by now," Barbara says, "and he kept
telling us that if the tree went, he'd grab the kids and
hang onto the cyclone fence. But we all knew that if the
tree fell, we were done for."

They made a strange sight, an entire family standing
among the scraggly branches of a pear tree, above a black
ocean of moving water. The minutes passed, the children
shivering in the icy wind, and still the tree held. Two

more bodies floated by. Michael whimpered and hung on his mother's arm. Finally, after what seemed to be more than half an hour, the water receded and they climbed down.

The wreckage, the bodies, the freezing wind, the black water, the desperate minutes stranded high in a pear tree, Barbara Adkins doesn't like to remember them: "Most of the time, I blank it all out," she says.

The flood left twelve people dead in Stowe, and took out all but five of the thirty estimated homes which once stood there. The water smashed through the Island Creek Coal Company's tipple office. And it was especially devastating for Stowe Bottom resident Taylor Hammons, who watched his eleven cars and two trucks hurtle to destruction. ("I used to trade and traffic in them old cars," Taylor said later, "just like a lotta boys trade coon dogs. But that water wiped me out for sure.")

The water moved more slowly now, as it surged out of Stowe toward a string of tiny settlements—Crites, Latrobe, Robinette, Amherstdale—about halfway down the hollow. Most of those who would die on this terrible Saturday had already perished. But now there was another hazard: pushing tons of wreckage before it, the water began backing up at the railroad and highway bridges, forming lakes which would expand until the pressure became too great. Then the bridges collapsed, and the water and wreckage moved on.

"It just terrified me," said Emma Workman, a middle-aged, plump and nervous woman who fled the wave moments before it struck in Crites. "I said, Lord have mercy! And I went to crying. I couldn't help it. There was a big rumbling going on, a rumbling and a roaring noise, like there was something under the water."

Emma Workman had been up only a short time when

the flood surged through her town. "We had us a pot of coffee and was a-drinking. My brother [Glen] was still in bed. Then the lights flickered off and on, and we saw the power lines shaking. I said to my husband, 'I wonder what's making those lines go? They're waving back and forth!'

"He said, 'Oh, they're probably working on 'em.' "

A moment later, Emma left her husband, Melvin, at the kitchen table and walked into the bathroom. Suddenly she heard a neighbor shouting at her family to run for the nearest hill. Peering through the bathroom window, Emma saw the wave. "Ah, it was terrible. I saw some houses in upper Crites go and then a big building. It looked like an apartment house. [This was probably the large, concrete-walled activities building at the Buffalo Boy Scout Camp, which was knocked from its foundations and washed downstream.] It was terrible. When you see homes going, you know all them people don't get out."

Emma and Melvin grabbed their coats. For a few, terrifying moments, brother Glen refused to get out of bed. Melvin shouted, "You better get out quick, or that water's gonna take you away!" Glen did. The Workmans scurried across the railroad tracks near their home, to the safety of a hillside. "It stays on your nerves," Emma said later, "you don't just forget it. It has a bearing on you. 'Course, I try to keep from thinking about it, but you know, it doesn't die down overnight. I thank God that I didn't lose no loved ones in it."

Leaving Crites, the water moved on into Latrobe, where it demolished half of the twenty-odd homes in that small settlement, and killed three people. Next in the path of the floodcrest was Robinette: another twenty-five-or-so houses were destroyed, and three more lives ended when coal miner David Gunnels's young family was trapped inside its house trailer. The damage would be less severe now, as the wave's force lessened with distance. But dozens of families were yet to watch the destruction of their homes. And at least one more Buffalo Creek resident —an elderly invalid trapped on his sickbed—would die.

CHAPTER 7

Braeholm

At Braeholm, three miles below Robinette, seventy-two-year-old Willard Adkins had been up for three hours when the disaster struck.

Small and wiry, a sharp-jawed blade of a man, Willard Adkins needed no alarm clock to summon him from sleep. Not after his forty-nine years, eight months, and twenty-seven days in the coal mines of West Virginia. Not after the decades of arising before dawn each morning to make his early, underground shift. After all those years, Willard knew that getting up at 6 A.M. would be a habit for the rest of his life.

Now he yawned, and stretched, and made his way through the darkened house to the kitchen. First, a cup of black Chase and Sanborn percolated coffee, strong enough to "Get up and walk!" Then a bowl of hot Quaker Oats with milk and sugar. He sat in the quiet kitchen, sipping the coffee, spooning up the oats. It was like a thousand other Saturday mornings in the seventy-two years of Willard Adkins's life.

The son of a rural farmer, Willard had gone into the West Virginia coal mines as a fourteen-year-old boy in 1913. Those were the primitive years, in which the miners —without a union to assist or protect them—worked ten-to-twelve-hour shifts in gloomy, death-dealing tunnels for $1.25 a day. The work was backbreaking: first the miner knelt at the face of the coal seam, turning a double-crank breast auger in order to drill a hole for dynamite. Then he placed the powder inside the hole, packed it tightly in place with a tamping stick, and lit the fuse. If the resulting explosion didn't bring the roof down, the miner of that day picked up his No. 4 coal shovel and spent the next six or seven hours loading the "shot" coal into small railroad cars.

Willard began his career as a "trapper boy," which meant that his job was to open and close the trap doors which kept the ventilation system working as men and coal moved through various sections of the mine. After a few years, the hard-working Adkins moved up to loading coal—at one cent per bushel. By hustling, he could turn out 125 bushels on a shift and take home $1.25 at the end of the day.

"It was plenty rough in them days," Willard says. "They didn't have any machines like they do now, you had to do everything by hand. . . . When that dynamite went, it cracked about three times louder than a shotgun shell, and you could see the shale fall. I've seen lots of it fall—like fifty and sixty carloads at a time." The shale (or slate) falls were dangerous, but not nearly as feared as the sudden plunge of a treacherous kettlebottom. "Those are the worst," Willard says, "because they don't pop nor crack nor nothing, and you can't tell when they're going. I've seen men mashed all to pieces with 'em. When they'd get it, they'd just come and get 'em and the undertaker would take 'em out for burial."

In those early days, the coal company controlled every part of the miner's life. The workers were paid on the basis of production; they were often cheated at the scales, and the fear of "getting shorted" in the company pay office never left them. Once paid, the miners bought their groceries at a company store—using currency (or "scrip") also printed by the company—at sky-high prices. They lived in ramshackle, woodclap houses rented from the coal operator. The system was designed, in short, to get the maximum production out of each worker, while providing him with the minimum necessities to keep alive. Workers who protested these conditions were quickly fired, and then permanently blacklisted by the other coal operators in the area. Zealous clergymen who preached reform found their congregations dwindling and their collection baskets empty. Although there were individual instances in which coal owners went out of their way to help the miners, most old-timers today remember that era as a kind of slavery.

Willard Adkins, for example, remembers an incident at a mine on nearby Pond Fork in 1926. After a long, grueling shift in which he had loaded six or seven three-ton cars with coal, he walked down to the pay office and asked the company bookkeeper for two dollars' worth of scrip so that he could buy groceries that night. Adkins expected no argument; he knew he had earned several dollars' worth of the company currency during the preceding few days. But the bookkeeper, a surly man in a worse-than-usual mood on that particular day, refused.

"He turns around and tells me I ain't got two dollars' worth of scrip," Adkins recalls. "He says, 'You ain't got it in there. You ain't loaded enough coal to fire my cooking stove!'

"I says, 'Well, I guess I'll have it in a little bit.'

"I went home and got my pistol and I came back. I laid that .38 special barrel up there in the window and I says, 'Now gimme five dollars' worth of scrip!'

"He says, 'Yes, sir! Yes, sir!' He gave me that scrip all right, and he never mentioned a word about it to anybody! He knew I had it coming."

There wasn't a lot of money during those days, but Willard says you didn't need a lot: "Why, you could live on $5 a week. Bacon was ten cents a pound, lard was four cents a pound, you could buy beans at a cent a pound. You'd buy a whole barrel of flour at once, and make biscuits out of it for a couple of months. Really, when you came down to it, we ate pretty good."

Adkins spent a lot of hours in the mines, of course, but he also found time for leisure. He simply stopped sleeping. "I wasn't so tired," he says with a squinting chuckle. "Why, in them days, I was much of a man. I'd walk five miles to the mine in the morning, and then five miles back home at night. Then I'd go out to a dance and stay all night—or maybe go coon hunting with some of the other boys. Yes sir, I was much of a man."

By the late 1920s, two new developments were rapidly changing the world of the southern West Virginia coal miner. The first was the introduction of heavy machinery into the mining process—a labor-saving innovation which immensely reduced the grinding load of drudgery which the miners had been carrying. By 1930, in fact, Willard Adkins was driving a huge electric cutting machine—as well as making a heady eleven cents a ton in wages.

The second development was not so pleasant, but it would do more in the long run to boost pay scales and improve working conditions than the miners had ever dreamed possible. During the 1920s and 1930s, as the militant United Mine Workers labor union gained strength under the fire-eating John L. Lewis, the coalfields of southern West Virginia came under increasing pressure from outside organizers. Coal-rich Logan County was at the heart of the conflict, and became the focus of countless efforts by union men to break the owners' stranglehold on the entire region.

Adkins remembers the company guards (rented deputy sheriffs, usually) who patroled the portals at every mine during that era. He remembers the company spies whose job it was to inform on fellow-miners who demonstrated union sympathy. As the fight between the owners and the unions escalated, violence in the Logan coalfields became a daily experience.

"Why, it was a regular war," Willard says. "Both sides had their own armies. They had guns and ammunition; they was fighting battles all across them mountains. They'd load the dead on box cars, just like you was loading ties on a flat-bed railroad car! You had to watch everything you said, or you'd wind up shot."

Adkins survived the labor wars by keeping his mouth closed and getting his work done. Lean and tough, his instinct for survival pulled him through a dozen different brushes with death. The worst accident took place on the 27th of January, 1957, when a compressed air blasting device (called an "air dock" by the miners) exploded in Willard's arms as he was attempting to repair it. The air dock, which by the 1950s had replaced dynamite as a method for shooting coal in many mines, is basically a gun. Powered by thousands of pounds of compressed air, it shoots a jet strong enough to shatter tons of coal instantly. The 18,000-pound explosion blew Adkins's safety helmet apart, shattered his glasses and punctured one eardrum. "It was just like a gun going off," he says. "My face and eyes looked like beefsteak that had been pounded. It blowed all the hide right off my face, and it knocked me down for awhile. But I went ahead and done my shift's work." Incredibly, Adkins lingered on to finish his shift before departing for the hospital—and beginning a one-week absence which the doctors ordered after seeing his injury.

The years of violence, the years of unending work were over for Willard Adkins now. The five children which he and his wife Grace had raised had long ago left home to establish families of their own: three of them having settled in Michigan, one in Ohio, while son Ralph had chosen to stay behind in West Virginia. Living with sixty-year-old Grace, a quiet, graying, grandmotherly woman in their pleasant, paid-for house at 267 Braeholm, Willard spent his retirement days puttering around on odd-jobs, watching television and reading the family Bible after supper each night.

Now it was seven o'clock. Grace still slept quietly in the bedroom; Willard decided to wander outside and take another look at the creek. All day Friday, people had been predicting that the little stream was about to come over its banks. Willard wanted to see for himself. Strolling back into the bedroom about an hour later, Adkins told his just-awakening wife that Buffalo Creek looked perfectly safe. He didn't think there would be any problem whatsoever.

After sleeping awhile longer, Grace got up and headed out to the kitchen to fix her own breakfast. She was just settling down to a cup of coffee when a yellow-painted, Amherst Coal Company truck skidded to a stop next to the Adkins's home. As they stepped to the door, the man inside the truck shouted, "Get out of your house! The dam's broke, and Lundale's gone!"

"We had about ten minutes' notice," Grace says. "I didn't have time to grab anything, not the bank book or my pocketbook or anything. I just threw a housecoat over my duster, and we run."

Running now, the Adkins scurried a few feet down the hollow road to where Willard had parked their 1969 black Ford pickup truck. They jumped in the truck with the water already swirling around them, and Willard stepped on the gas. They drove down the hollow about twenty yards and then turned left in an attempt to cross the valley floor and reach safety on a hillside. They were too late: the surging water—still more than ten feet deep in places—hit the truck broadside and swept it downstream.

Spinning and lurching, the truck swirled in the racing tide while they sat helpless. Finally, it came to a stop—wedged under the front porch of a house which had broken loose from its foundations but was backed up against a large tree.

They were trapped. They sat in the truck, wedged and powerless, while the water rose inch by inch. Grace remembers a crowd of people standing on the opposite bank, watching them. "We were yelling at the people," she said later, " 'Help us! Help us!' But there was no way they could get to us. There was nothing they could do. Finally, they just turned away and went back in their houses. They didn't want to watch us drown."

They watched as the water crept up the sides of the truck. Soon it was lapping against the windows, seeping slowly through the frames. A few more minutes, and the flood had covered the truck entirely. They were underwater. They prayed.

"I said, Lord, You've divided the waters before, You can save us if it's Your holy will," Grace remembers. "We was just talking to the Lord," Willard said. "I asked the Lord to help me, and he did. When it went down, I thanked the Lord."

After what the Adkins said seemed like thirty minutes, the water receded. Grace knew her prayer had been answered. "I said, 'Lord, cause this water to go down,' and it did. It was the Lord that saved us." The Adkins had lost their home and their truck, but their lives had been spared. Grace talked about that moment, crying quietly as she remembered it. "Everybody looked good to me after that. I was so glad to be alive. Seems like this flood has drawed people close together. Everybody looks good to everybody now."

Willard Adkins, after his forty-nine years in the coal mines and his brush with death in the Buffalo Creek flood, spoke more harshly than his wife: "The coal companies, they're the cause of it all," Willard said. "They can't turn a wheel with me."

"It was real gloomy," James Miller said later, "a dreary-looking morning. I saw my brother-in-law's car coming up the back alley, and the thought came to me that somebody was sick or else he wouldn't be out that early. I kept on with the coffee, and then there was a knock. It was my brother-in-law's son, telling us to get out of the house because the dam had gone."

James Miller, a skinny, middle-aged coal miner who lived with his wife, Tilda, and his four kids at Braeholm, was suddenly in a dilemma. What should he do with the sick man, the invalid, who lay at that moment on his specially-rigged hospital bed in the back room? James Brunty, Tilda's 83-year-old father, was already close to death. A black lung victim who had suffered three strokes, Brunty lay in a perpetual coma, with tubes running into his throat, his arm, and his gall bladder. Moving him out of the house would be extremely difficult and might immediately send the old man into a lethal attack of pneumonia.

It was an excruciating dilemma. "I said, 'Well, Christ, I can't just leave the house!'" Miller remembers. "I knew if we took him out of bed, away from the suction machinery [used to clear phlegm from the old man's throat] and the warmth of the house, he would catch pneumonia and die. And we'd had a lot of other false alarms up there. Every time you got some rain, somebody would holler, 'The dam's gonna break!'"

James Miller decided to do what he could. First, he moved the family car to high ground, and put his four sons (ranging in age from eight to fifteen) inside. Then he walked out to the creek, trying to estimate how bad the damage would be when the floodwaters came. Tilda, meanwhile, was making frantic telephone calls. The man at the Logan radio station told her his office was checking out a rumor that the dam had broken, but they had nothing definite on it yet. She hung up and dialed the number for the Man Police Department. The man on the desk told her the police were also checking out several reports of a flood in Buffalo Creek Hollow. He suggested she call the local National Guard headquarters for further information. "That's when I really panicked," Tilda says. "I couldn't understand why he'd send me to the National Guard if he didn't know something was wrong."

A few minutes later, walking down the hollow road outside his home, James came across Logan County Deputy Sheriff Max Doty, one of the two deputies who had been dispatched to evacuate the residents of Buffalo Creek hours earlier. Miller asked Doty how bad the danger was; the deputy told him that "seven or eight homes have been torn up in Lorado."

Doty didn't seem to think the damage would be very severe in Braeholm. But he took off in a hurry, interrupting Miller's questions with a quick, "Sorry, I gotta go." Returning to his home, James told Tilda that the dam had indeed broken, but there was a good chance that the Miller house could withstand the flood. The minutes passed, and they sat. "You could hear voices begin to raise and people begin to scatter out of the way." Miller looked through a kitchen window at the creek. It looked "real black, raising up." Foot by foot, the water climbed up out of the stream bed, until it was level with the road. Increasingly nervous, James walked outside. He stood staring at the water, talking to a neighbor, a strip mining inspector from the state's Department of Natural Resources named Richard Frazier. "There isn't enough water up there to do much damage," Frazier told Miller, "I just checked that yesterday." (Frazier and several other inspectors from the West Virginia DNR had spent most of Friday reviewing a Buffalo Mining Company strip mine permit application which involved 1,000 acres of land near Three Forks.)

The water rose, and soon it was running against the side of the house. The situation had grown critical. Miller realized that the invalid would now have to be evacuated from his home. He shouted for help. Almost immediately, two employees from the local water company, who had been doing some utility work nearby, joined Frazier and Miller inside the threatened house.

Shouting at Tilda and her mother to join the children in the car, Miller and his hastily-assembled rescue team turned their attention to the sick man. As he gathered up a set of pillows on which the invalid would be placed, James heard a chilling shout: "Here she comes!" A moment later, the two frightened utility men "broke and ran.

"They told me, 'I'm sorry, Buddy, I gotta go, I can't stay,'" James remembers, "and then they just ran away. I told Richard, I said, 'You go to your family—I'll look after the old man,' and he took off. By that time, the water was already breaking over the Braeholm Post Office, and it was knee-deep in the front yard." Miller turned back to his father-in-law. "I began gathering up all the bedclothes and pillows, everything I could get ahold of and stacking them under him, raising him up. I was hoping that if the water did get in the house I could control the bed and maybe make it float for awhile and then set back down.

"Then we got the large impact of it and the houses began to shift and pile up and disintegrate and crumble." Miller's home was tearing apart around him. "There came a point," he remembers, "where I had to try to save myself. I had to decide when I would leave him to save my own life."

Struggling out to the front of the collapsing house, Miller attempted to climb up on the roof. But as he stepped on the front porch railing, the whirling water tore the entire porch loose and flung him face-down in the flood. Thrashing blindly in the water, he latched onto a telephone pole and for a moment thought himself safe. He wasn't. The pole was broken off at the bottom, and hung suspended in the flood by the wires attached to it. In a few seconds, a clump of passing wreckage caught in the wires and pulled the pole out of his arms. "There was a guy wire attached to the pole," James says, "it had gotten between my legs and I couldn't break loose. I had juice going right through me, and I couldn't get loose of that pole. I didn't get the full impact of the line—if I had, I'd have been cooked right there in the water. I wasn't scared at first, because I was a pretty good swimmer. Not until I got into that juice. When I couldn't get loose from it, I thought, 'Oh, boy, you're a goner now! If you get an impact, a surge on that thing, you're gone!'"

But Miller was lucky. "I heard a noise. It turned out to be a transformer blowing. I looked over at the mouth of Ruffner Hollow there, and I could see the smoke and fire coming from that transformer. Then the power shut off. I picked out a tree down there at lower Braeholm, and began fighting my way over there. Someone had driven a big nail into the side of the tree. I got a hold on it until I could catch my breath, and then I climbed out."

Safe for the moment, Miller was frightened that his wife and sons—after watching him get swept away in the flood—might have plunged into the water in a foolish attempt to help. The fears proved groundless. Tilda and the kids had driven the car safely out of danger, and all of them were soon re-united. It was some solace, at least. Tilda's father, probably never knowing what hit him, was washed hundreds of feet down the hollow. All the evidence indicates that he was the last person to die in the Buffalo Creek flood.

In addition, the Millers' home had vanished without a trace. "We never found a board off the house that we could identify," James said. "We did find the refrigerator later—down in Accoville—and a cabinet or two, lying along the creek bank. Oh, and we found the hospital bed after awhile, too.

James and Tilda Miller were angriest about the lack of warning. "They'd been watching that dam since midnight," James said. "They knew it was in trouble all the time. If they'd just leveled with us, I don't think anyone would have had to die. We could have called an ambulance, and had him taken to a hospital. We could have done a lot of things, if only we'd known."

Before reaching Braeholm the flood had shattered about half of the seventy-odd homes in Amherstdale. Particularly hard hit was the area of Proctor Bottom, an exposed, lowland spot nestled between the creek and the hillside above it on the north side of the hollow. More than twenty homes were destroyed in Proctor Bottom, making an unhappy prophet of the U.S. Congressman who had said, after a gob slide struck the area in 1967, that the people at Proctor were "living under the gun of threatened annihilation."

"It was just pure hell, that's the only way you can describe anything like this," said Marvin McDaniels, a coal miner who lived in Amherstdale. "We saw some people go by in a mobile home [probably David Gunnels and his family], floating right down the middle of the hollow. Never did find out if they got out."

Another Amherstdale resident, Douglas Harvey, climbed up on a mountainside to escape the wave. "I seen water stretched from mountainside to mountainside, and buddy,

that was a lotta water!" Harvey's father-in-law, Claude
Rogers, also escaped at the last moment. "I saw houses
start crowding in on each other, so many houses you
couldn't even see the water behind them. You can just
imagine houses trying to climb over top of each other with
a big wall of water behind 'em. They hit the crossing,
about 100 yards below us, and some of 'em started breaking. The others just climbed up on the trestle."

For Oat Bailey and his wife, Annabelle, the last-minute
escape from Amherstdale was almost comical. While Annabelle loaded sixteen children, her daughter-in-law, and an
uncomfortable dog into the family's 1971 Ford LTD station wagon, Oat stood around laughing, scoffing at the
very idea of an approaching flood. Annabelle and the kids
had the last laugh, however—as Oat scrambled furiously
up a hillside to avoid the torrent. "He just laughed," Annabelle chuckled later, "just stood there and laughed at us.
And then he got a busted mouth, getting himself out
later!"

The water ripped into tiny Becco next, blasting a
grocery store, a barber shop, and three or four of the
fifteen houses in the community. Backing up behind the
Becco railroad trestle, the water-driven wreckage formed
a grotesque, forty-foot-high pile. Later, National Guardsmen, equipped with bulldozers, would dig there for days,
as the search for bodies continued. Just below Becco, in
a sparsely-settled area called Riley, fourteen-year-old
Rocky Graham sat on a hillside rock and watched the
disaster unfold. "It looked horrible," Rocky says, "all this
black mud coming down throwing everything, washing
houses down. It was throwing big water tanks around, and
all kinds of stuff. I didn't know what to think. I just sat
and watched it all wash away. It just busted, that's all I
know." The flood erased a few more homes at Riley, and
then plummeted into Braeholm, where James Miller and

Willard Adkins would struggle with their separate fates.

Another Braeholm resident, a normally soft-spoken widow named Faye Maynard, spotted the debris coming down the creek and drove to safety in her convertible. "I shouted, 'Let's get out of here!' Faye said later, "and we jumped in the car. I said, 'Lord, I left my door open—and a table full of dirty dishes!' I didn't realize what was happening, I didn't even look back. It happened that quick."

The worst was over now, as the spent floodwaters rippled through Accoville, Crown, and Kistler on their way to the Guyandotte River at Man. A few dozen additional houses and trailers were wiped out—most of them around Accoville—and a few more people were injured as the water dissipated into the big river at the bottom of the hollow. During its two-hour rampage, the runaway water had destroyed a total of 507 homes, according to a later study by the U.S. Department of the Interior. Another 273 homes had suffered major damage, and 663 more homes would need minor repairs. All told, about half of the homes between Three Forks and Man had been rendered uninhabitable, leaving perhaps 4,000 people without a place to sleep. Also lost were about thirty trailers, 600 automobiles, and at least thirty business establishments. The Chesapeake and Ohio Railroad line was extensively damaged, ten highway and railroad bridges were knocked out, and the telephone, sewer, water, and power systems lay in a shambles. Within a few weeks, the federal authorities estimated the total damage at more than $50 million.

The human toll was equally staggering. The dead numbered 125 (seven of the bodies have never been found), with an estimated 1,000 injuries of one kind or another. No one could estimate the psychological impact of the disaster, or count the number of children who would wake up with terrifying nightmares, or the grown men who would later quake at the mere sight of water.

Numbers and statistics will never tell the story of what happened to Buffalo Creek Hollow on that dreadful Saturday morning in February of 1972. From one end of the seventeen-mile-long valley to the other, Buffalo Creek looked like it had been bombed. In every direction, shattered homes lay tumbled like the pieces in some insane jigsaw puzzle. Amid such total destruction, little things seemed to stand out: a knife and fork, glinting side by side in the mud; a jar of pickled green beans swept from somebody's pantry; bloodstains running down the side of a brand-new washing machine, lying on its side in the muck.

The survivors say they will never forget it.

Investigations

CHAPTER 8

Rescue

THE VALLEY lay in shock,
pulverized as if by some enormous fist, and for a time only
the wind seemed to move. In silence, the drifting snow-
flakes settled on the piles of wreckage, on the bodies of the
dead. The victims lay crushed against bridges, wedged
between railroad cars, swinging grotesquely from tree
branches. In silence, a pale sun rose higher in the sky and
the winter breeze went looping through.

And then the valley began to stir. A child wailed for her
lost parents. A dog barked in quick, flute-like echoes, as
he nosed in the wrecked foundations of what had once
been his home. A wife dissolved in her suddenly-discovered
husband's arms. From Three Forks at the top of the hollow
to Kistler at the bottom, the cries for help—for food and
blankets and medical supplies and transportation and a
thousand other things—began.

Hour by hour, the reports of widespread destruction
had been filtering out of Buffalo Creek. By mid-afternoon,
the national news wires had begun running cautious
stories about the disaster. The first estimates were that
perhaps two dozen residents had perished. As time passed
and the list of fatalities grew longer, AP and UPI upped
the reported death toll to 50, to 75, finally to more than
100. The governor issued a statement calling the flood a
"major disaster" and asking the President for emergency
assistance. It was not long in arriving. The U.S. Army
Corps of Engineers stepped in to supervise the cleanup of

wreckage and the demolition of homes and buildings which had been rendered uninhabitable. The President's Office of Emergency Preparedness dispatched its regional staff to the hollow, with authority to spend up to $20 million on disaster relief. The Department of Housing and Urban Development sent representatives to begin planning a system of temporary trailer parks in which the homeless could be housed. (Within three months, HUD would build thirteen such parks, housing more than 2,500 refugees in 700 government trailers.)

For the moment, however, the first task was to evacuate the dead and injured. Various local rescue squads, accompanied by National Guardsmen equipped with bulldozers and other heavy equipment, had arrived in Buffalo Creek Hollow by Sunday morning. Town by town, they combed through the wreckage for bodies. Many of the victims, buried under tons of debris, were not found for several days.

The evacuation of the injured was slow. Most of the roads in and out of the hollow were blocked, and the main Buffalo Creek highway was completely shattered in most places. The nearby coal companies threw their radio-equipped trucks into the battle, and after a few hours, the first company and Civil Defense helicopters arrived on the scene. Many of the injured rode out on huge National Guard trucks, which could plow through the heaviest mud and cross the shallow creeks without using a bridge.

The rescue effort was conducted on a massive scale. Officials arrived from the West Virginia State Police, the state Department of Highways, the National Guard, the Army Engineers, the West Virginia Employment Office, the Southern West Virginia Regional Health Council, and a dozen other agencies to speed the evacuation and reconstruction. But it took days to organize the rescue—and every delay meant additional suffering for the victims, waiting for a ride to the hospital, a shot of penicillin, or an assurance that a loved one had been found safe.

As the death toll inched upward, the Logan County Sheriff's office and the State Police established a temporary morgue at a red-brick gradeschool in South Man. Soon the bodies of the dead, encased in black plastic zipper bags, were lined up in rows on the gradeschool floor. For days, the relatives and friends of the dead moved woodenly up and down the ranks of bodies.

The water had disfigured some of the victims beyond recognition, making the situation even more grotesque. In at least two cases, bodies were mistakenly signed over to supposed relatives, only to be returned later when the shocked mourners suddenly discovered their error. "It was the strangest thing I ever saw," said Amherst Coal Company clerk Mearl Bartram, "Their faces would swell up from the water until you couldn't tell who was who. I must have looked at my father-in-law five times without recognizing him. Finally I identified him by the metal pin in his leg."

While the National Guardsmen dug through tons of smoldering wreckage and the Salvation Army served hot meals up and down the hollow, some of the rage and resentment which had been fueled by the disaster finally flashed to the surface. Surprisingly, however, the cries of outrage came not from Buffalo Creek Hollow—where most people maintained a cool, wait-and-see attitude—but from Charleston, the state capitol.

Infuriated by what he described as "irresponsible reporting," West Virginia Governor Arch A. Moore suddenly lashed out at press coverage of the disaster. Within a few days, the situation had escalated into one of the murkiest —and bitterest—disputes between media and government on record. Moore, a portly, deep-jowled, silver-haired Republican usually identified with coal company interests, was angered by published reports in which a Buffalo Mining Company official seemed to be blaming the state of West Virginia for the disaster. Both the Louisville (Ky.) Courier-Journal and the Associated Press carried stories in which Buffalo Mining Superintendent Ben Tudor suggested that the state's anti-pollution laws were partially to blame for the flood. Noting that the state had consistently denied Buffalo Mining permission to drain off its massive impoundment, Tudor was quoted as saying, "They were too concerned about the trout downstream. It either had to be the people or the trout, and now both are gone."

Angered, Moore charged that the reporters who were quoting Tudor had failed to check with the state before running their stories. Had they only done their homework, the governor insisted, the writers would quickly have learned that Buffalo Creek was not a trout stream and that it had never been stocked with trout by the state's Department of Natural Resources.

The governor had additional proof of journalistic irresponsibility: he played a tape of a telephone conversation in which a Pittston Company (Buffalo Mining's parent-corporation) spokesman assured the governor that Tudor had been misquoted. The Pittston official explained that what Tudor had really said was, "A few years ago it was industry practice that when a sludge dam got full of water during heavy rains it was released into the stream. The regulatory agencies no longer permit this water and this black water to go into the stream and all . . . water must now be retained within the impoundment."

Some observers felt that the corrected quotes just didn't sound like they'd come from the mouth of a mine superintendent. And the controversy became more tangled a few days later, when the governor was apparently proved wrong. The state of West Virginia had indeed been stocking Buffalo Creek with trout during past years. It said so, right on the trout-stocking schedule printed annually by the Department of Natural Resources. Increasingly irritated by the reports, however, Moore had already stepped up his attack by declaring the devastated hollow off-limits to the press.

Now the conflict reached laughable proportions. Denied access to the hollow, newsmen from across the country began complaining about suppression and censorship. A national television network, after having one of its camera crews turned back by the State Police, blasted the governor for his high-handed tactics. The ongoing brouhaha prompted an amusing, if apocryphal, story: a TV news reporter approaches a tall West Virginia state trooper at the mouth of Buffalo Creek Hollow. He is denied permission to enter.

"Look," the reporter argues, "Walter Cronkite sent me here to get a story. What do I go back to New York and tell Mr. Cronkite?"

The trooper replies, "You tell Walter Cronkite that Corporal Garrett said 'No!'"

After a one-day blackout, Governor Moore relented and the hollow was again opened to the press. But as far as the governor was concerned, the damage had already been done. Describing the press coverage of the entire disaster, the governor made this incredible statement: "The only real sad part about it [the coverage] is that the state of West Virginia took a terrible beating which far overshadowed the beating that the individuals that lost their lives took, and I consider this an even greater tragedy than the accident itself."

If Governor Moore was conspicuous by his continuing barrages against the press, The Pittston Company—which many thought had ultimate responsibility for both the dam and the disaster it triggered—attracted increasing attention by its unbroken silence. Day after day, as the death toll mounted and the terrible extent of the tragedy became obvious to the entire nation, Pittston's corporate New York offices remained silent, without even a word of condolence for the victims of the flood. Only once during the two weeks which followed the disaster did a Pittston official have anything to say about it.

But when he finally spoke, it was a bombshell: "We're investigating the damage," a Pittston spokesman told Charleston Gazette reporter Mary Walton in a telephone interview, "which was caused by the flood which we believe, of course, was an Act of God." The spokesman went on to tell reporter Walton that there was nothing wrong with the massive gob pile at Three Forks, except that it was "incapable of holding the water God poured into it."

Instantly, the phrase "Act of God" became the most controversial words spoken during the entire disaster. Hundreds of hollow residents, devoutly attached to their fundamentalist, God-fearing religion, deeply resented the suggestion that God had been responsible for the disaster. "They're trying to blame it on God," was a comment heard again and again in Buffalo Creek Hollow, "but God didn't pile that slate up in the hollow. Men did." For many, the classic response to Pittston's argument was offered by a an elderly lady who supposedly stood up at a neighborhood meeting, addressed the corporate lawyer who had just blamed the flood on God, and said, "Tell me, Sonny, you ever see God riding a bulldozer?"

After the "Act of God" incident, Pittston remained silent until March 9, when the corporate headquarters issued a statement to the company's shareholders:

We are sure that you have heard or read about the catastrophe which struck in West Virginia on February 26. On that day more than 100 persons lost their lives and many were injured in a flood when a porous impoundment used as a water filtration system by the Company's subsidiary Buffalo Mining Company in Logan County gave way.

No amount of sympathy can help those who lost their lives, but the Company and each of its subsidiaries in the area are making available aid and assistance to the survivors. Everyone on the scene went to work as soon as possible to effect rescue operations. All available personnel were summoned and communications via portable radio transmitters were established with areas outside of Lorado. All available equipment was rushed over the mountains, and our helicopters were immediately put to work in rescue operations. . . .

We wish to express heartfelt sympathy to the families of all who have suffered loss as a result of this tragedy. We will continue to help the flood victims and work with all those engaged in assisting the people of Logan County.

The help was not long in coming. Although Pittston believed "the investigations of the tragedy have not progressed to the point where it is possible to assess responsibility," the big conglomerate on March 28 gave its subsidiary, Buffalo Mining, the go-ahead to open a disaster claims office in Man. Soon the claims for damages were pouring in by the hundreds.

Day by day, the rescue work continued. The bulldozers dragged down the last of the wrecked buildings, and soon huge, smoke-billowing piles of burning debris lined the hollow. The last of the dead were buried, and the first HUD trailers were arriving to shelter the homeless. (Many of them lived for weeks on temporary cots in hallways and classrooms at the Man High School.)

It had been a massive project—with the Red Cross administering hundreds of typhoid immunization shots, with various relief agencies bringing in tons of blankets and clothing, with the Salvation Army serving almost

400,000 cups of coffee during its three weeks on the scene. All in all, more than 2,000 workers from dozens of different organizations had combined their efforts to pull Buffalo Creek Hollow out of the mud. Things might never return to normal, but at least they were under control. Now there would be time to investigate the disaster, to study the causes and the people responsible for so much anguish.

CHAPTER 9

Inquiry

Within a week of the Buffalo Creek Flood, the U.S. Department of the Interior announced that it was launching a full-scale investigation into the disaster. The order came from the office of Hollis M. Dole, assistant Interior Secretary for Mineral Resources, in a memo to both the U.S. Geological Survey and the U.S. Bureau of Mines.

Dole asked the agencies to "establish immediately a joint Bureau of Mines-Geological Survey Task Force to study and analyze hazards associated with the disposal and storage of coal mine waste materials, with initial focus on the retaining banks in the Appalachian coalfields." The Task Force was assigned to analyze the Three Forks dam failure as well as to identify other hazardous coal piles which might be expected to give way in the future.

The Task Force—a quickly-assembled team of geologists, hydrologists, and mining engineers—departed immediately for Buffalo Creek. The Geological Survey sent twenty-two technicians to the dam site, where they collected soil samples, took aerial photographs, surveyed the now-empty reservoir behind the failed dam, and ran a "flood profile" on the entire valley.

One of the first surprising discoveries made by the Geological Survey team was that the flood could not be blamed on the weather. In fact, the late-winter rainstorm (Pittston's "Act of God") which preceded the dam's collapse had not really been unusual at all. Statistics provided by the U.S. Weather Bureau showed that the storm dumped 3.7 inches of rain around Buffalo Creek in the three days before the flood—an amount of rainfall which could be expected to occur at least once every two years in southwestern West Virginia.

The weather report said: "Streams similar to Buffalo Creek in and around Logan County responded to the three days of precipitation with flows on the order of a ten-year flood; that is, a flow that can be expected to occur on the average of about once in a ten-year period. Following the failure of the coal-waste dam, flow in Buffalo Creek near Saunders [Three Forks] jumped from less than a ten-year flood to a discharge about forty times greater than a fifty-year flood. The difference between a discharge less than a ten-year flood and the discharge 40 times greater than the fifty-year flood reflects the difference between the natural flood that would probably have occurred and the flood that resulted from the failure of the dam." The evidence was compellingly clear: the weather could not be held logically accountable for either the break in the dam or the flood which followed it.

The Geological Survey study pointed out that the Buffalo Creek impoundment had not been designed to serve as a water-retaining dam: ". . . banks No. 2 and No. 3 were not engineered as dams and would not be acceptable as dams in an engineering sense. . . ." Then it speculated about the actual cause of the break: "Failure of the coal-waste dam probably occurred through foundation deficiencies, causing sliding and slumping of the front face of the dam. The failure was accelerated by the waterlogged condition of the dam. The slumping lowered the top of the coal-waste dam and allowed the impounded water to breach and then rapidly erode the crest of the dam."

The thirty-two-page Geological Survey study next described the process by which the monster dam had been constructed: "In constructing dam No. 3, the coal waste was dumped from trucks in closely spaced piles from four to seven feet high and then graded in layers two to four feet thick. The dumping was carried across the dam in the form of 'lifts' on successive levels of material ten to twenty feet thick. . . . Trees in the path of the dam construction were not removed but were covered by dumping. The pool area also was not cleared of vegetation. The sludge on which the waste was dumped was only partially displaced and much of it formed the foundation of the dams."

After noting that the collapsed embankment had contained no "open or overfall type of spillway" (a basic safety feature on most dams), and that its single, twenty-four-inch overflow pipe was too small to have been effective, the geologists proceeded to a minute-by-minute description of the break. Written in cold-blooded engineering language, the narrative still had a chilling effect:

After 7:30 A.M., the top of dam No. 3 was cut by cracks extending from abutment to abutment parallel to the faces. Water was rising to the crest through these fractures and other holes. The front of the dam was sliding off and the crest was lowering noticeably. Apparently by this time pore pressure had passed the maximum, the flow line within the bank was near the crest, and the entire right side of the dam was buoyant and was being driven to the downstream face. The pore pressure was further relieved by internal slippage and slumping that finally produced total failure through overtopping of the slumped blocks.

Total failure occurred about 7:59 A.M. when the right side of the dam breached along a line starting about 120 feet from the front of the right abutment and trending diagonally towards the valley wall at the rear. Extreme turbulence threw coal-laden water 300 feet from the dam and splattered cars on the haul road.

. . . Within seconds, dam No. 2 was topped . . . the clear pool was filled, overtopped, and the small dike on its right side breached, and the shed and transformer pole was destroyed at 8 A.M. . . . The torrent of water crossed the haul road at the south end of the burning coal-waste No. 1. . . . At the lower end of the burning coal-waste bank, the water increased the hydrostatic pressure within the bank, causing explosions of steam and producer-type gas. The three or four explosions reported were severe enough to shake the ground at Saunders and raise mushroom-shaped clouds of ash and smoke. Moments later, at about 8:01 A.M., the torrent of water entered Buffalo Creek. The cascading torrent followed an existing depression within the coal-waste bank at the front, which curved abruptly to the west. This abrupt curve diverted part of the initial surge towards the church at the mouth of Lee Fork

The geologists discounted several alternate theories which had been advanced to explain the break in the dam and then listed the five causes of failure upon which they agreed:

1. The dam was not designed or constructed to withstand the potential head of water that could be impounded. It was primarily a waste pile that grew from routine dumping of waste.

2. Spillway and other adequate water-level controls were not in the dam, and no provision had been made for removing water once it had entered the pool behind the dam. The capacity of the 24-inch pipe was too small to handle large flows, and the pipe was placed so high that water rising to it greatly increased instability of the dam.

3. The sludge on which the dam was placed was inadequate as a foundation. Seepage through the foundation gave rise to extensive removal of material (piping). The weak foundations also gave rise to ex-

tensive slumps and subsidence which led to the initial breach and overtopping.

4. The great thickness of the dam (from front to back) in relation to height without engineered compaction led to a decrease rather than an increase in the stability of the dam. The thickness, along with some relatively small compaction, impeded seepage through the bank and thus produced a higher phreatic surface. The high phreatic surface was reflected in saturation of the dam and the resulting buoyancy accelerated the failure. The impediments to initial drainage also limited the effectiveness of seepage in regulating the height of the impoundment.

5. The dam was constructed of coal waste, including fine coal, shale, clay, and mine rubbish. This material disintegrates rapidly, is high in soluble sulfates which reduce bonding strength, is noncohesive, and does not compact uniformly. A safe, economical dam could not be constructed from such material alone.

In short, the government geologists were using their technical language to say what any layman could easily understand: Buffalo Mining had constructed an enormous dam out of garbage by dumping tons of unstable material on a treacherous, coal-sludge foundation. Moreover, the un-engineered barrier had been thrown together without any of the obvious safety features (such as spillways) which any responsible engineer would have automatically included in the design. The implications were clear: the Three Forks dam, as designed, constructed, and maintained, had simply been doomed from the start.

Meanwhile, the Department of the Interior had commissioned another study of the cause of the dam failure. Performed by Fred C. Walker, a civil engineer at Interior's Bureau of Reclamation, the analysis differed little from the one assembled by the geologists. Walker added one additional factor to the long list of possible causes:

he suggested that "The outer part of the downstream surface may have been frozen so that seepage water could not escape and the pressure built up behind this layer until it blew out."

Walker was equally critical of the manner in which the Three Forks dam, and others like it, had been built:

> Locally these barriers are called 'dams' but to me this is unacceptable nomenclature. These structures were created by persons completely unfamiliar with dam design, construction and materials, and by construction methods that are completely unacceptable to engineers specializing in dam design. Although the state of West Virginia requires permits, approval of plans, and inspection during construction for impoundments more than ten feet deep, I was unable to find out that such requirements had ever been complied with. It was inferred that these impoundments were initiated at the instigation of the Department of Natural Resources, to clarify waste water from coal processing operations to improve the water in the streams so that they would support game fish.

Near the end of his nine-page report, Walker introduced a theme which was to become increasingly popular with the various government agencies investigating the flood. Pointing out that the hazardous impoundment had been built in order to prevent coal-wash waste water from polluting West Virginia creeks, Walker said, "I suspect that there was promotion of pollution control practices without a thorough evaluation of the side effects of the procedures adopted that produced these conditions and it must share some of the responsibility for this disaster."

One page later, in another apparent reference to the state's anti-pollution laws, Walker concluded his report by saying: "Responsibility for this disaster cannot be placed entirely on any one individual or organization; there were many contributors." The implication was an interesting one. Overzealous West Virginia legislators, in their concern for developing stringent anti-pollution measures, had apparently contributed to the disaster. By prohibiting the mining company from further pollution of the state's watercourses, the lawmakers had forced Buffalo Mining to build a water-retaining dam. Later, according to this logic, the collapse of the hazardous, shoddily-built dam proved that the anti-pollution laws were short-sighted and badly needed overhauling.

If Walker's logic was puzzling to some, the stance which the Department of Interior took during the next few weeks almost defied understanding. First, the Department announced that it had no jurisdiction—no authority—to regulate coal-waste dumps like the one which failed at Three Forks. Almost at the same time, however, the Department's Bureau of Mines launched a gigantic inspection of all such gob piles in the Appalachian coalfields. The Bureau identified more than 900 of the dumps—immediately classing some of them lethally dangerous—then began issuing violations against the offending coal companies. (There was no protest over this activity by an agency which supposedly lacked the authority for it).

As the evidence of the Buffalo Creek dam's shoddy construction and maintenance continued to flow into Interior's Washington offices, the Department suddenly discovered another cause of the disaster: failure on the part of Buffalo Creek residents to flee the area in time. Interior found this information important enough to base a press release on. Dated March 15, the release was headlined: BUFFALO CREEK RESIDENTS HAD PRIOR WARNING OF DISASTER, INTERIOR OFFICIAL TESTIFIES. It read in part:

Residents of Buffalo Creek hollow had been warned of an impending disaster which did not occur on at least four occasions before the February 26 dam collapse and flood that claimed at least 116 lives, an Interior Department official testified today.

"As a result of previous false alarms, the residents were reluctant to move when the dam did fail," Hollis M. Dole, Assistant Secretary of the Interior for Mineral Resources, told the House Interior and Insular Affairs Subcommittee.

There are reports that warnings were given, that those who responded immediately were able to reach higher ground above the flood, but those who hesitated were lost," Secretary Dole said.

Some observers wondered exactly what purpose the Department of Interior release served; others questioned its accuracy, since it was common knowledge that dozens of the victims had been sound asleep in bed when the water struck.

Meanwhile, the political heat generated by the disaster had become stifling. While U.S. Senators and Congressmen took pot-shots at the Bureau of Mines for its failure to prevent the disaster, Bureau spokesmen defended themselves by insisting that they were not authorized—or even equipped—to regulate coal-waste dams like the one at Three Forks. Appearing before the U.S. Senate Labor Subcommittee, Elburt F. Osborn, the harried director of the Bureau of Mines, introduced his remarks this way:

"I appear this morning in response to your invitation of April 24, 1972, to assist in your investigation of the disaster that occurred on February 26, at Buffalo Creek, West Virginia.

"This occasion is both solemn and serious. Your inquiry into the terrible tragedy of Buffalo Creek, while essential, is not likely to inspire joy in any of us.

"It can, however, inspire hope. And from what I have been told of your committee's investigations thus far, I am hopeful. The members of your staff, in their contacts with the Department, have been courteous and diligent. They have had to ask many questions, but those questions have been asked, not in a spirit of witch-hunting, but one of truth-seeking."

Convinced that the Senate Labor Subcommittee would do no "witch-hunting," Osborn still felt compelled to construct a long, complicated argument which sought to prove, in essence, that the Bureau of Mines could not be blamed for the disaster:

Among these mandatory safety standards [created by the Federal Coal Mine Health and Safety Act of 1969] are requirements for the maintenance and inspection of surface installations, including retaining dams. These requirements essentially impose an obligation on the operator of a coal mine to construct, inspect, and maintain retaining dams in a manner which will not create a hazard, and which will prevent injuries to men working in the mine. It is crucial to emphasize at this point that a comprehensive review of the legislative history of the Federal Coal Mine Health and Safety Act of 1969 clearly demonstrates that the specific thrust of the Act is to protect the health and safety of miners, meaning men working in a coal mine. In practical terms, and keeping in mind that persons expert in the construction of dams would not have known that the dam would fail until almost immediately prior to its collapse, even if a Bureau coal mine inspector had been at the dam site as the water rose, he could only have found the existence of an imminent danger to the miners working on the mine property. This would have resulted in an order withdrawing the miners from the area of the mine affected by the imminent danger or prohibiting the miners from entering the area. The finding by a Bureau coal mine inspector of an imminent danger and a subsequent withdrawal order, however, would not have prevented the retaining dam from failing. Furthermore, the inspector's order would not have applied to persons off the mine property in the path of the flood; and, therefore, the mine operator would not have been legally bound by the order to remove those persons from the flood path.

Very basically, the Federal Coal Mine Health and Safety Act of 1969 does not give the Bureau of Mines the legal authority to protect the public from hazards arising in a coal mine. Without additional statutory authority, mandatory safety standards regarding retaining dams on coal mine property, and enforcement of these standards can only apply to the prevention of hazards to miners on the property.

Osborn was admitting, in short, that his agency had the power to inspect coal-waste dams. The problem was, he said, that the Bureau of Mines only had authority to protect *miners* from hazards created by a coal company. Therefore, the argument went, the Bureau of Mines could not have been expected to protect the public from the disaster.

Perhaps not. But the 1969 law contains at least one crystal-clear provision—Section 77.216:

(a) If failure of a water or silt retaining dam will create a hazard, it shall be of substantial construction and shall be inspected at least once each week.

(b) Weekly inspections conducted pursuant to paragraph (a) of this S. 77.216 shall be reported and the report shall be countersigned by any of the persons listed in paragraph (d) of S. 77.1713.

The record spoke for itself: the Buffalo Mining Company had not obeyed either regulation. No inspection reports on the Three Forks dam had ever been filed with the Bureau of Mines (in fact, the Bureau cited the company for this violation after the disaster). Osborn's agency, it seemed obvious, had failed to enforce one of its laws. Enforcement, if properly applied, might have caused the Buffalo Mining Company to correct the problem—or it might have led to a complete shut-down of the company's operation, and a further investigation.

The Senate Labor Subcommittee was not satisfied with Osborn's explanation. Reacting angrily to two days of testimony from flood survivors, Subcommittee Chairman Harrison A. Williams of New Jersey blasted the bureaucratic negligence which he said had contributed to the tragedy:

> As I said yesterday, we have got to stop the next disaster in its tracks. I find it unconscionable that government agencies at all levels are quibbling and nit-picking about the scope of their jurisdiction while these "man made" disasters continue to take the lives of innocent citizens.

> I will say to the Bureau of Mines that three years ago we gave you a job to do, we gave you responsibility, we gave you authority, and we gave you the funds to do that job. We, as elected representatives of an outraged public want it done now. Other laws may be useful and may resolve the ivory tower hypothetical problems of your legal purists, but, only your action, now, can avoid another disaster.

> We will close these hearings now, but not our concern with disasters. Since Buffalo Creek, another "man made" disaster struck at the Sunshine Silver Mines claiming ninety-one lives. This too lies within the jurisdiction of the Department of the Interior, and we are going to bring to light the causes of that tragedy for public scrutiny.

Senator Williams was not the only critic of the U.S. Department of the Interior in the weeks which followed the Buffalo Creek flood. Dissenters charged that the Department—controlled by a pro-business, Republican administration—was taking too soft a line toward the giant coal industry. They noted, for example, that The Pittston Company had been assessed more than $1.3 million in safety violations during the year which preceded the flood, but that the company, by appealing every single one of the fines to the limit, had escaped so far with less than $300 in penalties. These critics thought they saw a connection between the fact that the Secretary of the Interior's brother—Thruston Morton—sat on the board of directors for The Pittston Company, and the tolerant, chummy stance which they thought the Department was taking toward the coal giant responsible for the flood. Under Interior Secretary Rogers C. B. Morton, the dissenters charged, the Department would continue to go easy on the huge coal producers which dominate the industry today.

As for Elburt Osborn, the embattled director of the U.S. Bureau of Mines, the disaster and its aftermath had been a most unhappy experience. Sitting in his Washington office on a steamy summer day, months after the killer flood, Osborn shook his head: "They're looking for a scapegoat," he said sadly. "These things happen, and they look around for somebody to blame them on. I guess the federal government is a good scapegoat, and that's what we've become."

On March 8, the U.S. Senate Labor Subcommittee announced that it was also conducting an investigation into the Buffalo Creek flood disaster. At the Subcommittee's request, the U.S. Army Corps of Engineers began assembling an in-depth analysis of the entire flood much like the one prepared by the Interior Department's Task Force. The Army Engineers ran laboratory tests on soil samples, took photographs, and surveyed the disaster area. Their conclusions added up to an indictment of the Buffalo Mining Company operation:

The basic concept of dam No. 3 was not acceptable from an engineering standpoint. This concept, as understood from reports, was for the dam to serve as a retention structure for a pool of relatively shallow depth for settling coal washings. The key point is that the success of the operation depended on seepage from the pool through the embankment and/or foundation to clarify the "black water." No other provision was made for passage of water except the pipe at an upper elevation which had never been in use until the February 1972 storm occurred, and, it is understood, was not intended for use during normal operations. Also, no design and construction effort was made to alleviate possible detrimental effects of the required seepage. We then have a structure the successful operation of which depended on uncontrolled seepage; unless some happy accident occurred whereby Mother Nature took care of this fundamental error of conception, the dam was doomed to failure from the time the first load of refuse was dumped. Many points of design and construction on these three dams are considered inadequate in varying degrees; these are:

a. Inadequate by-pass system for high volume flows.
b. Lack of proper measures to assure adequate foundation.
c. Lack of zoning in dam and other measures to assure control of seepage.
d. Lack of compaction.
e. Lack of erosion protection.
f. Little attention to steepness of embankment slopes.
g. Lack of qualified technical inspection.
h. Little attention to the preparation of the abutments prior to embankment construction.

> i. Continued dumping of embankment material on poor material (sludge).

The 225-page Army Corps of Engineers report became the cornerstone on which the Senate Labor Subcommittee hearings were based. Conducted in late May, the two-day Senate probe listened to a variety of witnesses from Buffalo Creek as well as to representatives from the corporate offices of The Pittston Company in New York. The senators heard pathetic stories from West Virginians who had survived the flood, and they listened to company explanations of how the dam had been built and failed. When the hearings ended, Senator Harrison Williams summarized them:

"Yesterday, all of us here spent a full day reliving one of the worst disasters in the history of the State of West Virginia. We heard the heart-rending account of the survivors and we learned why the Buffalo Creek dams should never have been built.

"When I asked the company what engineering design was used, I was shocked—frankly. All they did was to start dumping junk on top of sludge and called it a dam. And how did they reinforce the dam when told that it needed reinforcement? They just dumped some more refuse and slate on top of an already failing structure. This was as stable as a child's sandpile on the shores of an ocean and reflects as much engineering judgment.

"We have also learned that there are numerous other such dams, some in a potentially imminent state of collapse."

In addition to its analysis of the failed Buffalo Creek impoundment, the Army Engineers studied several similar structures in southern West Virginia. Their findings made ominous reading:

The reconnaissance of the area during selection of the dams for study and the study per se, indicated no dams with conditions identical to those on the Middle Fork —there may be some—they simply were not observed. Some are similar, with only minor differences; most, however, were significantly different—with better foundation conditions and generally better in method of construction. Although different, none of the dams observed are considered safe structures; they were not designed and constructed to withstand the natural destructive action to which they can be exposed. Since the dams surveyed all show evidence of some prior failure and the dams studied are believed to be representative of those in the area, it is reasonable to assume that sometime during their construction almost all of the other dams in the area have failed, with varying degrees of gravity. None, yet, have failed with the catastrophic consequences of those on the Middle Fork. However, contemplation of the appalling possible results in the area of a deluge normally associated with a hurricane of major proportions or of a moderately severe earthquake in the early spring when the embankments would be more nearly saturated than at other seasons, is stupifying, in that the recent Buffalo Creek disaster could be multiplied manyfold.

The Interior Department and Senate Labor Subcommittee investigations ended with all parties agreeing on one point: new legislation was needed immediately to eliminate the possibility of future disasters like the one at Buffalo Creek. The Interior Department, in particular, recommended speedy passage of the Administration's pending "Mined Area Protection Bill," a comprehensive set of laws which supposedly dealt directly with coal industry problems like strip mining, gob piles, and coal-waste dams. Swift passage of the new legislation would give the Interior Department authority to clear up the problem once and for all.

Environmentalists and other critics of the government agencies were not convinced, however. Many of them felt they were seeing the same old pattern repeat itself: first a terrible disaster, then the urgent pleas for new legislation, then another disaster, followed by inevitable complaints that the regulatory agencies involved lacked the authority, or the funding, to take action. The critics noted that, except for the small fine attached to the Bureau of Mines' citation of the Buffalo Mining Company for failure to inspect its dam, no federal agency had in any way penalized Pittston, or its subsidiary, for its negligence in a disaster which killed 125 people. Without stiff penalties— or perhaps even the possibility of criminal prosecution— many observers wondered if the profit-oriented industrial giants would ever feel compelled to spend the time and money required to eliminate hazards like the one on Buffalo Creek.

On March 2, the state of West Virginia launched its own investigation into the disaster as Governor Arch A. Moore announced the formation of a special "ad hoc commission" which he said would study the killer flood from top to bottom. The composition of the nine-member panel immediately provoked questions, however, with critics charging that the commissioners appointed were in no way representative of the public-at-large. Headed by J. H. Kelley, the dean of the West Virginia School of Mines, the panel also included four West Virginia state officials, a representative of the U.S. Bureau of Mines, a member of the U.S. Geological Survey, a newspaper editor, and an industrialist. Fearing a whitewash, dissenters quickly pointed out that all four state officials (the director of the Department of Mines, the director of the Department of Natural Resources, the director of the Public Service Commission, and a state geologist) were directly involved in the disaster, as members of the state's various coal industry regulating agencies. The two U.S. Department of the Interior members would labor under the same handicap, the critics argued. In addition, they viewed Chairman Kelley as one of the strongest pro-coal spokesmen in the state. Charlie Hylton's Logan Banner was seen as coal's biggest backer among West Virginia newspapers, and few critics thought the industrialist—who depended on coal to run his business—very likely to make waves. The investigative panel included none of the Buffalo Creek survivors. It included no coal miners. In fact, the dissenters charged, it included nobody who made less than $15,000 a year.

Governor Moore weathered the criticism, however, and the panel held the first of its several sessions in late April. The pattern of the hearings was soon established: they consisted of polite question-and-answer exchanges in which no witness was ever pressed. Lacking subpoena power, the commission was forced to rely exclusively on voluntary testimony. Its questioning was a mish-mash of hazy speculation and friendly chatter. On the few occasions when one of the commissioners questioned a witness closely, Chairman Kelley stepped in to cool off the proceedings. At one point, for example, as Commissioner Hylton pressed a Pittston Company engineer named Richard Yates, Kelley butted in to produce this startling series of exchanges:

Q. (by Commissioner Hylton) Mr. Yates . . . It looks to me like that sometime, at some level of engineering, either in Dante [Virginia, Pittston's area headquarters] or Buffalo or New York or some place, that an engineer, in going over the maps of Middle Fork, would maybe take a double look at that map and say, 'Here, wait a minute. We have a hollow a mile long . . . And we are building an impoundment there that is going to be forty-three feet high . . . And behind that dam we are going to impound seventeen or twenty million cubic feet of water, and if that thing ever breaks, we are going to have a problem. . . .' It seems like some engineering person, somewhere, would have caught that.

A. (by Yates) That's right.

Q. (by Hylton) I just can't understand why sometime or other they didn't strike the idea, looking at the map, even if they had to take the double take . . .

A. (by Yates) I can't annswer that. I wasn't there long enough . . . (Chairman Kelley interrupts.)

A. (by Kelley) Commissioner Hylton, you know, none of us are perfect.

A. (by Hylton) That is true, I grant that.

A. (by Kelley) There have been very famous bridges collapsed very shortly after they were built . . . (the questioning turns to something else).

The flavor of the proceedings was perfectly demonstrated during a session in which Chairman Kelley refused to permit Pittston's overall safety record to be introduced into testimony. Explaining his refusal, Kelley argued:

> In the judgment of the Chairman, an equally responsible agent is the society itself which allows for the canalization of public pressures on various single-minded problems of the day to the sacrifice of more objective problems. Here is a condition of great danger to modern society, which condition, if allowed to continue, will surely lead to future disasters and increasing shackles on modern life.
>
> To waste the time of the Commission in establishing court-type evidence on one of the responsible parties, which party is obviously and admittedly involved in the disaster, simply denies the Commission the time and

effort to delve into some of the less obvious causes and origins of the flood. On this basis, then, the matter of The Pittston Company's compliance with health and safety laws is far less germane to the task of the Commission than are the multiplicity of causes and origins with which the Commission is charged.

Commission critics found this a peculiar use of logic. The panel had been convened to seek out the causes of the flood, but Chairman Kelley, in the middle of the hearings, had already concluded that "Canalization of public pressures" (presumably, the state's anti-pollution laws, which led to the dam's construction) was one of the causes. In other words, Kelley had already reached his conclusions and was introducing them in the middle of the investigation to prove that Pittston's safety record was not germane to the inquiry.

Another high point in the commission hearings came when Steve Dasovich, the Buffalo Mining vice president who had been in charge of the dam that morning, was asked about his confrontation with the two sheriff's deputies near the Lorado schoolhouse. Dasovich denied telling the deputies (both of whom quoted him later) that his men had "dug a ditch," or "channeled around" the threatened dam. "The only thing I recall telling Mutters," Dasovich said, "was that, 'Otto, you can go on home now. Everything is o.k.' Something to that effect. But I didn't discuss any technology with him."

Asked about the lack of engineering at the Three Forks dam, Dasovich said, "Well sir, to me that is common practice throughout the coal mining regions." And then later, asked for his opinion as to the cause of the dam failure, Dasovich confessed to being completely mystified: "Well, sir, I have thought about that thing for three months now, and I just don't know. I can come up with a dozen different ideas on it."

Typical of the commission's concern with conversation rather than facts was this fuzzy exchange between Buffalo Mining President I. C. Spotte and Commissioner Hylton. Promising at first, the questioning finally petered out as Hylton failed to press the witness:

> Q. (by Hylton) Mr. Spotte . . . There is a law in the West Virginia code which requires that a permit be obtained from the Public Service Commission for any dam ten feet or higher that obstructs a water course. Do you know whether or not the Buffalo Mining Company obtained such a permit before constructing dam No. 3?
>
> A. No sir. We did not. Although we did not consider, Mr. Hylton, that a dam. It was an impoundment, or embankment.
>
> Q. But it still blocked a water course . . .
>
> A. It blocked a water course to a certain extent. Yes sir. (The questioning turns to something else.)

Perhaps the most dramatic—and unusual—moment during the long series of hearings came when Wayne Brady Hatfield came thundering into the high school gym where that day's session was being held and challenged anyone present to a "fair fist-fight." While the commissioners sat tight-lipped and uncomfortable at this lack of proper decorum, Hatfield got it all off his chest:

> A. (by Hatfield) I'd like for every one of them to be tried and sentenced to murder, just like I took a gun and killed somebody they would try me for murder. Every one of them are guilty and they should be tried for murder . . . they should be tried for murder because they're guilty of murder, the coal companies. The coal companies said in the paper "What the hell do we care?" yesterday evening, "We've got our insurance paid off." What do they care for those people's lives? I challenge him [Dasovich] right here if he thinks he can make me out a liar. Let him walk up here and tell

me because he's talking to an old man that ain't scared. Besides that I'm looking for Steve Dasovich. Let them send me to the penitentiary if they want to. I've got nothing, not one thing. I've lost my wife and daughter. I lost one of the wonderfulest women that ever lived on earth and I'm looking for Steve Dasovich. If he's going to stay he better go do something about it and get out of the way because I'm looking for him.

Q. (by Chairman Kelley) All right, what else do you think?

A. (by Hatfield) If this country is dirty enough and low-down enough to put me in jail for what I've said, I just hope and pray to God to have mercy on the people that are dirty enough to put me in jail because I'm saying I'm going to get Steve Dasovich. I worked for him at the mines and he's the dumbest, low-downest man that ever walked on the face of the earth and I want everybody to know it. I want that coal company to know it. He didn't have sense enough to know that when he cut that ditch through there and dynamited it, or whatever he done to it, that every bit of it was going to come at one time . . . Anybody that's dumb enough to set and wait for it to come over the top and not think that when it comes over the top that the water won't go all at once—he's a fool. He doesn't have a brain in his head. He's a crazy man. . . .

The United States is getting low enough to let these companies kill and destroy people any way they want to. If a man gets killed in the mines and it's the company's fault, they'll say it's the man's fault and that's the truth. I heard them do it. If I worked in the mines 100 years I wouldn't tell a lie for the coal mines. That's all the coal companies are, dirty, low-down liars anyway they go at it. They don't care who they kill just so they make a dollar and that's all I've got to say.

Published in early September, the ad hoc commission's report surprised most of the critics, who found its recommendations and conclusions to be fairly hard-nosed, in

light of the earlier testimony. In frank language, the commissioners blamed The Pittston Company for the disaster:
"The [dam] failure . . . was solely the cause of the Buffalo Creek flood. No evidence of an Act of God was found
by the commission." The report went on to assert that the
company's actions deserved public censure: "The flagrant
disregard shown by The Pittston Company and the coal
industry for the safety of persons living on Buffalo Creek
and others who live near coal refuse impoundments
should be publically recognized. . . . The Pittston Company, through its officials, has shown flagrant disregard
for the safety of residents of Buffalo Creek and other persons who live near coal refuse impoundments. This attitude appears to be prevalent throughout much of the
coal industry."

The strongest language in the lengthy commission report came in the "Conclusions" section, where the nine-
member panel hinted that some of its witnesses might
have lied and suggested that a Grand Jury might be able
to do a better job:

> The commission has not been able to resolve ade-
> quately some of the conflicting testimony. Nor has it
> been able to corroborate some of the testimony. All of
> the opinions, conclusions, and recommendations were
> made in light of what appears to be the weight of the
> evidence as voluntarily presented to the commission.
> We are unable to assess many parts of it, since some of
> the witnesses called chose not to appear. We could not
> press this issue, since we were without subpoena power.
> It should also be noted that this commission has no
> authority to prosecute for perjury or contempt, and
> therefore, the witnesses, although under oath, had no
> fear of subsequent legal action for statements which
> were intentionally misleading or untruthful.
> We recommend that the proper judicial authority

with subpoena power . . . determine if Grand Jury or other appropriate legal action should be taken in order to more closely scrutinize testimony.

A few days after the release of the report, a Logan County Circuit Judge took the commissioners at their word, and ordered a Grand Jury probe into the disaster. The jury, which was scheduled to begin its investigation in November, was charged with determining whether or not criminal charges should be brought against officials at both Buffalo Mining and Pittston. In addition, the judge barred Logan County Prosecutor Oval Damron and his two assistants from the proceedings, because of possible conflict-of-interest (all three attorneys were representing flood victims in private suits at the time).

The governor's ad hoc commission—which looked to most people like a whitewash from beginning to end— had, in fact, introduced the possibility that those responsible for the flood would one day stand trial for it. And most observers, familiar with West Virginia's traditional, cozy stance toward the coal companies, saw that as a sign of progress.

While the various government investigations continued, a group of concerned West Virginians formed their own, private committee to assess the causes and origins of the killer flood. Called "The Citizens Committee to Investigate the Buffalo Creek Disaster," the twenty-member panel collected the testimony of eyewitnesses, studied the history of past failures at the Buffalo Mining complex and assembled all of the government engineering data it could find. Charging that Governor Moore's ad hoc commission was "Mainly a whitewash, packed with people who had a personal interest in the findings," the Citizens Committee included residents of Buffalo Creek, coal miners, union men, and church ministers among its members. Their conclusions differed sharply from the ones reached by the various state and federal agencies:

"Our conclusion is that the Buffalo Mining Company committed murder," said committee member Norman Williams, a former top official in the West Virginia state government who had earlier resigned to protest his department's tolerant stance toward strip mining. "This was a case of murder, clear and simple. The company had plenty of evidence [the two earlier, less-extensive failures, in 1967 and 1971] that the dam was unsafe. They must have been aware of the danger. But they took no preventive measures at all. They had no system to warn the public. Their first concern was profits, period. They were grossly negligent, and their negligence killed 125 people. We define that as murder."

By September, Williams's committee had assembled reams of evidence in support of its conclusion. Within a few weeks, the committee planned to take its findings to a Logan County Grand Jury and seek criminal indictments against several company officials for their part in the disaster. Williams had little hope that a Logan County jury could be convinced to indict the powerful coal men, but thought the effort might at least serve to focus public attention on the facts which his group had collected.

For the most part, the press investigations into the causes of the tragedy were shallow and fleeting. While most of the larger American newspapers sent reporters to the scene during the first days of the disaster, all but a few quickly dropped the story after it lost its initial impact. Two exceptions were the Detroit Free Press, which compiled an extensive, in-depth analysis of the flood (with particular emphasis on the lack of warning) and the Washington Monthly Magazine, which ran a lengthy study of the dam's history and The Pittston Company's involvement in the disaster.

In spite of the thousands of pages of testimony, and the

countless engineering diagrams, many questions about the actual sequence of events at the dam on that terrible morning will probably never be answered. There is strong —if not conclusive—evidence, for example, that, contrary to its later testimony, the Buffalo Mining Company did in fact alter the surface of the dam in the hours before it broke.

The evidence includes testimony from hollow residents who talked to a local bulldozer operator after the disaster. Several of them quote him as saying he had been engaged in digging a spillway in the dam's surface shortly before its collapse. The evidence includes an eyewitness who saw a length of pipe lying on the face of the dam— inside what looked like an open ditch—around 7:30 A.M. that morning. It includes Dasovich's own reported comments to the deputy sheriffs in Lorado: "We've dug a ditch," and "We've channeled around that thing." It includes several residents of Three Forks, who heard what they are sure was a bulldozer leaving the area of the dam soon after the break. These, and numerous other items of evidence, have convinced many Buffalo Creek residents that Buffalo Mining actually dug an emergency ditch in the dam—and perhaps triggered the flood.

In the end, however, the question of whether or not Buffalo Mining took last-minute steps to save its failing dam seems less important than several other, clearly-established facts: first, the un-engineered dam was in violation of two different laws; second, the company had ample evidence that the dam was unsafe; third, the company had provided no warning system of any kind to alert the hollow in the event of catastrophe; and fourth—perhaps most important of all—by persuading the sheriff's deputies to call off their rescue attempt, the company nullified the one, last-ditch effort which might have saved more than 100 lives.

By early fall, Pittston's subsidiary, Buffalo Mining, had

paid off hundreds of claims arising from the flood. Most victims settled for the state of West Virginia's statutory $10,000 award for "wrongful death," and then hammered out compromises with company adjusters on the value of the property they had lost. By then, Pittston had already issued a statement assuring its stockholders that "The ultimate effect of such claims [by victims] should not be material in relation to its [Pittston's] consolidated financial position." Most observers (including several attorneys then in the midst of settling claims with the company) agreed with this prediction. They felt sure the Buffalo Mining Company's liability insurance—estimated at $25 million—would be sufficient to pay for all of the claims, and they were convinced that The Pittston Company, which enjoyed net profits of $43 million in 1971, would not be required to pay out a single penny in settlements.

Later events may prove them wrong, however. In September, the prestigious Washington law firm of Arnold and Porter launched a $50 million lawsuit—on behalf of several dozen flood victims—against Pittston. Seeking both compensatory and exemplary damages, the plaintiffs' attorneys expected a long, uphill court fight before any damages could be paid. But they were confident they could "pierce the corporate shield" and force Pittston to share the financial responsibility for the disaster with its little subsidiary.

Perhaps the saddest finding to emerge from all these investigations was the one advanced from a government engineer, who estimated that The Pittston Company could have built a safe, stable dam at Three Forks for less than $200,000. Had the corporate giant been willing to sacrifice one-half of one percent of its 1971 profit, the horror of Buffalo Creek might never have happened.

CHAPTER 10

Response

One who would wish to understand the people of Buffalo Creek—and their response to the killer flood of 1972 —must first go to church with them.

"You can't be taken up with this world," says the fat lady in the billowing, rainbow-print dress. She sits on a folding chair in the sundazzled, woodframe church. She holds her Bible before her like a shield. She gazes far off through the windows, to the green-splashed hills of West Virginia. "If you take up with this world, you're just fooling yourself. You won't get anywhere. You won't feel right about things until you accept Jesus into your heart, until you feel that heavenly sorrow."

Heavenly sorrow. A man cannot be "saved" until he surrenders his heart—and his destiny—to Jesus. Once saved, he has earned heaven forever, regardless of what he does. It seems a somber, almost fatalistic view of things, a seemingly lifeless denial of human freedom, human free will—unless one considers the daily stresses on the average coal miner, the unending fear of death or injury with which he is required to live. Then the parts fall together; the religion makes more sense; once saved, death loses its terror. Once saved, the coal miner can face his daily shift secure in the knowledge of his ultimate, heavenly salvation. It is a way of dealing with the anxiety he has always known.

But this stern, fundamentalist faith also serves, at times, to inhibit the believers' capacity for action, to leave them with a passive, shrug-of-the-shoulders response to even the grossest injustices:

"God controls everything," said the man in the next pew, his face a somber, closed mask of piety. "I do believe you could call that flood an Act of God. Whatever happens, God wills it. If all those people got killed, it was because the Lord wanted it. Some folks have been saying there was a lot of sin up on Buffalo Creek Hollow. A whole lot of sin."

Most area residents refused to blame the Buffalo Creek disaster on "sin." But many of them, happy in their devout, fundamental faith, were willing to leave the entire matter in God's hands. "Everybody's gonna be judged," said Oldie Blankenship of Lorado, and "and the guilty's gonna be punished. There's no use anybody trying to take re-vengeance against something like this, because you can't better it no way. The Lord said, 'Vengeance is mine, and I will repay thee.' If it was their carelessness that killed the people, then they're gonna have to stand responsible for it with the Lord. No sir, I'm not gonna sue anybody."

Kate Waugh, a 43-year-old housewife who lost her husband and four of her six children at Stowe Bottom, feels the same way. Describing herself as a woman who "believes strongly in church and religion," Kate is sure that the Lord will settle one day with The Pittston Company. Meanwhile, she says she has struggled to understand the disaster: "I believe He has a reason for every death that occurs. I do believe He has control over life and death. But I wonder sometimes, so many of mine, all at one time. They say not to question His doing . . . but I do at times. You can't lose four children, a husband and a granddaughter and not question why. He knows what He did—but I don't know why He did it. I don't disagree . . . but you know . . . you do wonder."

Not everyone was so willing to trust in the Lord, however. For some Buffalo Creek residents, the disaster was a personal outrage which they would never forget. "I'll tell you my opinion of it," said Charlie Bevins, "and I'm a mine foreman, myself. It wasn't a thing but industrial murder. I mean, when you back up that much water on a slate dump, you ain't asking for nothing but trouble. . . . And the federal and state inspectors knew the condition that thing was in. Why wasn't something done about it? I been in this game too long. I know why. They line their pockets with money, the inspectors do."

Many youthful miners, like twenty-one-year-old Wilbert Hicks, Jr., saw the disaster as just another example of the big coal companies' lack of respect for both safety laws and other people in general. "They coulda fixed that dam if they wanted to," Hicks said after fleeing his home in Lorado. "But they didn't want to spend the time or money it would have taken. Listen, they do a lotta dumb things up there in them mines—I know, I work there. Anytime they can take a short cut on safety, they'll take it."

David Gunnels, the coal miner who lost his wife and two children at Robinette, agrees. "My wife and two kids was just as good as murdered by that coal company," Gunnels says. "But let me tell you the kind of thing they do. One time on my shift, the digging arm on the coal loader threw a rock up and knocked me out for an hour and a half. When I came to, the first thing the boss yelled was, 'Hey, can you run a loader?'

"I said, 'Brother, are you stupid enough to ask me if I can run a loader after me just getting hurt?' I said, 'Get me a mantrip [a ride out of the mine], I'll see you after vacation!' And I took a sixteen-day vacation, starting right then."

The most common complaint voiced about the Buffalo Mining Company's performance in the disaster dealt with its failure to warn the public about the dangerous dam. Paul Black, the balding, affable manager of the Island Creek Supermarket in Lundale, remembers that Steve Dasovich visited his store around 7:45 that morning. Dasovich told Black he wanted to by half a dozen raincoats "For the men who were working at the dam. But he didn't say anything about what they were doing up there. It didn't seem important at the time."

It became important a few minutes later, however. Black's wife, Betty, trapped in their home half a mile up the hollow, drowned as the flood surged through Lundale. Racing back to save her, Black had to leave his car and scramble for a hillside to save his own life. If Dasovich had only said something about the danger, Black figures his wife would be alive today.

Responding less vocally, some miners simply used common sense in assessing the responsibility for the disaster. "It was their business, wasn't it?" asked fifty-nine-year-old Allen Workman. "If I had damned that much water up and then turned it loose on people, I'd be blamed for it, wouldn't I? You can take it from there."

"They keep trying to say God caused this flood," charged Clarence Walls of Robinette, "but God didn't run a dump truck and pile all that slate up in the hollow."

Other miners were more cautious in their appraisal, anxious to look at both sides of the issue: "All I know is that the dam broke," said Buck Arnold, a chain operator for Island Creek. "I'm not gonna be a radical and jump out there and say it's the coal company's fault. You've gotta get both sides in the case before you start talking like that." Coal miner Johnny Wells thought the company had been criticized too much for the disaster: "They've modernized mining a lot, even since I've been here. You hear these old fellows talking about back when they used hand equipment. It was a lot more dangerous then. But

they got new laws now. As far as mining goes, they keep
it pretty safe anymore. They talk about the company a
lot—but you gotta figure their cost, too. If they didn't
make money, you wouldn't have a job."

Robert Albright, who lost most of his family in the flood,
insisted that the coal miners would have to share the responsibility for the disaster along with their bosses. "I
guess, in a way, we're all responsible. After all, we dug
that goddarn coal out of there, and that gob had to go
somewhere. In a way, I think there's just too many responsible. . . . There wouldn't be a jail big enough to hold
all of 'em that's responsible."

Barbara Adkins, whose family survived the flood by
climbing into a pear tree, suggested that anger would
only prove useless: "No, I can't say I'm mad at anybody.
I hated that it happened, I hated to see them get killed,
but being mad isn't going to bring anybody back."

Perhaps the strongest governmental response came from
West Virginia Congressman Ken Hechler, who had long
been a supporter of mine safety legislation and increased
black lung benefits for miners. "These people have been
living as prisoners of the coal industry," Hechler charged,
"and the officials at both the state and federal level have
been soft on the coal industry. The flood symbolizes the
trouble with the whole attitude toward coal, which has to
change if we're going to protect the people. We've got to
put emphasis on human beings instead of production and
profits. It's as simple as that."

Most of the officials at the Buffalo Mining Company
and The Pittston Company refused to talk to reporters in
the months which followed the flood. Jack Kent, the strip
mining superintendent who had been watching the huge
impoundment the night before it broke, was an excep-
tion. While he admitted that no engineering had been
done on the dam, Kent explained his company's failure

to warn the public by saying that Buffalo Mining simply
had no idea the water would prove as devastating as it
did. "Looking back at it now," Kent said, "I don't think
I'd change anything I did. I've got nothing on my con-
science at all."

The weeks passed, the gray winter receded, and soon
the mellow days of spring returned to Buffalo Creek. Beneath the gentle sunshine and the shuttling mountain
breeze, the tiny creek trickled on, peaceful and placid
as ever. In groups of two and three, the people returned
now to walk the winding hollow where they had once
lived. They lingered by the roadside, staring out over the
deep, grassy expanse where Lorado had stood. They
tramped the now-deserted flatland on which Three Forks
had once been located. And many of them began planning
for the day when they could move back to Buffalo Creek.

By now, the state of West Virginia had announced plans
for a new highway, a massive water and sewer project,
and a whole series of housing developments along the
hollow. Workmen were already hammering at the water-
wrecked Lundale Baptist Church, readying it for a new
congregation in the years ahead. Paul Black had repaired
and re-decorated his Island Creek Supermarket, and
would soon be re-opening for business. The emphasis was
on the future again. And everybody—from the state officials on down—was saying that Buffalo Creek Hollow
was going to be a mighty fine place to live in coming
years.

Some would not be able to forget the past, however,
or shed the recurring nightmare in which, like a great,
black ocean, the killer flood came roaring down upon
them. For Wayne Brady Hatfield, the past would be difficult to escape: "My nerves are gone. I don't eat but one
meal a day now, and I barely pick at that. And if I get

around water, I start to get real jumpy. I can't stand to get around that river, my nerves get so bad I almost go crazy. Even if it just rains a little bit, I start to shake."

And Freddy Church, a disabled miner who lived at Crites and who pulled a drowning woman from the water would have trouble forgetting. "It's on his nerves now," says Freddy's wife, Elizabeth. "He can't sleep at night. He keeps seeing people, their faces coming toward him in the water. A lot of people are like that—they're hurt mentally. I don't know if they'll ever get over it."

We will have to wait and see.